# The Integration Imperative

Michael P. Gillingham • Greg R. Halseth
Chris J. Johnson • Margot W. Parkes
Editors

# The Integration Imperative

## Cumulative Environmental, Community, and Health Effects of Multiple Natural Resource Developments

*Editors*
Michael P. Gillingham
University of Northern British Columbia
Prince George, BC, Canada

Greg R. Halseth
University of Northern British Columbia
Prince George, BC, Canada

Chris J. Johnson
University of Northern British Columbia
Prince George, BC, Canada

Margot W. Parkes
University of Northern British Columbia
Prince George, BC, Canada

ISBN 978-3-319-22122-9 ISBN 978-3-319-22123-6 (eBook)
DOI 10.1007/978-3-319-22123-6

Library of Congress Control Number: 2015954628

Springer Cham Heidelberg New York Dordrecht London


Printed on acid-free paper

Springer International Publishing AG Switzerland is part of Springer Science+Business Media (www.springer.com)

*Justice Thomas R. Berger*
*Who broke the mold and forged*
*a new direction*

# Preface

Throughout the first decade of the twenty-first century, we have witnessed unprecedented growth in the exploration and development of natural resources. Previously untouched landscapes are now the focus of government and industry, both racing to supply resources to growing, international economies, often without a full understanding of the implications of their actions. One need only look to western Canada, the central United States, or northern Australia to witness how the rapid extraction of petroleum and mineral reserves is outpacing our ability to measure and manage the corresponding environmental, health, and social impacts. For some locations, the development of natural resources has followed traditional patterns that have long supported communities. For other locations, development is a new and quickly evolving proposition, with numerous industries seeking a range of resources, from minerals and forest products to hydroelectric power. Those projects are superimposed on past land use and development activities, and the superimposition has implications, both positive and negative, for local ecosystems, people, and communities. The many consequences of multiple natural resource development projects are *cumulative impacts*.

The concept of cumulative impacts is not novel nor is it difficult to appreciate. The rancher struggling with shale-gas exploration and development and the community planner attempting to facilitate the growth of a transient workforce both understand the implications and complexity of the rapid development of natural resources. Since the 1970s, regulatory agencies from around the globe have recognized and struggled to identify, assess, and manage cumulative impacts, although largely from an environmental perspective. Despite that 40-year history, cumulative impacts remain one of the most pressing and complex challenges facing governments, industry, communities, and conservation and natural resource professionals. Although research and practice have improved our ability to assess and manage cumulative environmental impacts, much improvement is needed to account for the integration of communities and people within the cumulative impacts equation. Informed by our experiences in northern British Columbia (BC), a Canadian hotspot for these cumulative impacts, we have applied an integrative lens in this book to develop a broader and more holistic understanding of the full range of cumulative impacts that result from the development of natural resources.

This book responds to a series of long-standing demands and opportunities that are particularly timely in BC, but that are also relevant across Canada and internationally. Indeed, the needs of our communities echo a deep academic literature that has called for more holistic and ultimately effective approaches for assessing and managing cumulative impacts. For us, the local relevance and urgency of this challenge became obvious at a workshop held at the University of Northern British Columbia (UNBC) in January 2014, where we discussed the *Cumulative Environmental, Community and Health Effects of Multiple Natural Resource Developments in Northern British Columbia*. This event was hosted jointly by three UNBC research institutes—the Community Development Institute, Health Research Institute, and Natural Resources and Environmental Studies Institute—to provide a broad perspective on the development of natural resources. This was one of the first opportunities in BC to shift the discussion of resource development from a polarized and narrow focus on environmental impacts and economic opportunities to include a fuller set of challenges and solutions that encompass a wider range of cumulative impacts. The workshop involved a suite of focused presentations and facilitated discussions that were directed at generating ideas for maintaining ecological resilience and sustainable and healthy communities within the context of long-standing, but rapidly expanding natural resource activities. Participants came from both Aboriginal and non-Aboriginal communities, local and provincial governments, and the research and business sectors to meet, talk, and learn. This was a unique opportunity for cross-disciplinary and cross-sectoral dialogue that clarified the need to communicate, develop, and promote these ideas to a wide audience.

The ideas explored during the workshop underscored the timeliness of this issue and the demand for a broader perspective and more integrative approach to understanding the cumulative impacts of natural resource development. Informed by these concerns, and recognizing that the full range of impacts were disconnected and isolated in the policy, practice, and academic sectors, we focused on the challenge of integrating our multidisciplinary perspectives. Those early and heartfelt public discussions were the genesis of this book. Although the main authors provided research expertise in community development, public health, and environmental processes and change, we sought to bring a broader set of voices to the book. In the spirit of the complex and interconnected discussions that took place at the workshop, we invited colleagues to provide different perspectives on the concept of cumulative impacts. For some, cumulative impacts was a new term; however, given the breadth of the topic, their experiences in environmental, community, and health research and practice were highly relevant and added considerable insight. Based on that collective knowledge, both formal and informal, the book's chapters address a number of key challenges and potential solutions for addressing cumulative impacts:

- Understanding the cumulative impacts of natural resource development will require an integrative understanding of environmental, community, and health dynamics.
- Current regulatory approaches have a place in addressing the impacts of development, but we require a revolution in strategy, not an evolution of current practice, to develop an integrative framework capable of addressing cumulative impacts.

- An effective framework must be participatory and must involve all parties who create or are influenced by cumulative impacts.
- Cumulative impacts occur in a temporal context that connects past, present, and future impacts; thus, a strategic vision and long-term planning are essential.

Our collective experiences, both locally and internationally, suggest that these challenges and the implementation and testing of solutions will be relevant to researchers and students from a broad range of disciplines, as well as to policy makers and an educated lay public, including nongovernmental organizations. Although much of the text refers to natural resource development in BC, this provides examples of circumstances and specific challenges that are global in scope. The book has relevance to anyone whose work considers the implications of natural resource development for people and the environment.

The influence of the January 2014 workshop, the ongoing interactions among authors with diverse backgrounds and interests, and the scope of the challenge have resulted in a book with four interconnected themes that progress from a description of the problem to recommendations for addressing cumulative impacts at broad spatial and temporal scales. Part I of the book (Chaps. 1 and 2) includes two chapters that introduce the overall structure of the text, make a strong case for better assessment and decision-making processes, and provide a description of the essential terminology. The second part of the book (Chaps. 3–5) examines cumulative impacts from a multidisciplinary perspective, starting with the environment and progressing to a discussion of community and then health impacts. In Part III (Chap. 6), we present a series of vignettes that illustrate how these cumulative impacts shape and influence the environment, communities, and people. These contributions provide a human perspective on an often technical issue. This diversity of voices is supplemented by shorter inset boxes that are interspersed throughout Chaps. 3–5. The two chapters (Chaps. 7 and 8) in Part IV of the book revisit the limitations of the current philosophy and technical approaches for addressing cumulative impacts. We finish with a set of recommendations that are based on key principles and elements that could serve as an integrative framework for addressing the cumulative impacts of natural resource development at regional scales.

Both the origins and structure of the book represent a learning journey among a group of colleagues who are committed to addressing complex issues in a way that reflects our interrelated realities. Our hope is that the book will create a space in which others can re-think and re-envision our understanding of cumulative impacts, prioritize decisions that increase our options, and encourage a view of environmental, community, and health dynamics as parts of an integrated whole that will determine our collective future.

Prince George, BC, Canada

Michael P. Gillingham
Greg R. Halseth
Chris J. Johnson
Margot W. Parkes

# Acknowledgements

The impetus for this book arose from discussions at the workshop *Cumulative Environmental, Community and Health Effects of Multiple Natural Resource Developments in Northern British Columbia*, which was held at UNBC in January 2014. We thank the organizing committee of the workshop, Michael Gillingham, Greg Halseth, Bill McGill, Margot Parkes, Joanna Paterson, and Rachael Wells. We also thank the Community Development Institute, Health Research Institute, and Natural Resources and Environmental Studies Institute for their vision in identifying the need for this dialogue and hosting this event.

Funding for the event was provided by the BC Oil and Gas Commission. In response to a suggestion by UNBC's Vice-President External Relations, Robert Van Adrichem, the Commission came forward with a contribution to support a series of research projects that could help inform their staff about issues related to oil and natural gas development. Part of their funding supported the workshop that served as the genesis for this book.

Hosting these events at UNBC's Prince George campus was important and symbolic for at least two reasons. First, we recognized that the event was held on the traditional territory of the Lheidli T'enneh First Nation. We acknowledge this and give thanks for the stewardship the Lheidli T'enneh have shown for this land since time immemorial. We particularly thank Elder Edie Frederick for her opening prayers and thoughtful welcome that helped remind participants from all cultures that we have a collective responsibility to care for our environment and the communities it supports.

Second, we recognized that UNBC was created by the people and communities of northern BC just over 25 years ago. Part of UNBC's mandate is to deal with issues relevant to our region while also remaining true to the ideals of universities as a place for discussion, dialogue, and respectful debate on what are sometimes challenging, uncomfortable, or difficult topics. Thus, we felt that it was appropriate to talk about cumulative impacts at this time, in this setting, and at this institution.

We also thank Dr. Ranjana Bird, who at the time was UNBC's Vice-President of Research. Dr. Bird worked with the BC Oil and Gas Commission to secure their contribution to UNBC and committed to setting aside part of that contribution to allow UNBC's three Research Institutes to work collaboratively on this project. Her office has also provided support for the publication of this book.

The participation of Greg Halseth and Margot Parkes in the development of this book was supported by their Canada Research Chair appointments. Greg Halseth is a Canada Research Chair in Rural and Small Town Studies (CRC 950-222604) and Margot Parkes is a Canada Research Chair in Health, Ecosystems, and Society (CRC 950-230463).

As we began development of the manuscript, we wanted to remain true to the idea of widening the conversation around cumulative impacts. We sought a forum and format that would allow other voices to illustrate the breadth and complexity of the issue and help inform solutions. In the end, we decided to use inset boxes and vignettes from invited colleagues in the natural, social, and health sciences. Those who have joined us on this project have helped to ground, humanize, and extend our understanding of cumulative environmental, community, and health impacts. Brief introductions to these colleagues are provided in the Contributors section, but we also acknowledged their contributions here, along with their affiliations at the time of writing.

**Chapter 3**

Annie L. Booth, UNBC; Allan B. Costello, UNBC; Peter L. Jackson, UNBC; Donald G. Morgan, British Columbia Ministry of Environment; Justina C. Ray, Wildlife Conservation Society; Dale Seip, British Columbia Ministry of Environment

**Chapter 4**

David J. Connell, UNBC; Dawn Hemingway, UNBC; Derek O. Ingram, Dancing Raven Community, Lands, People consultants; Glen Schmidt, UNBC; Karyn Sharp, Dancing Raven Community, Lands, People consultants

**Chapter 5**

Henry G. Harder, UNBC; Dawn Hemingway, UNBC; Martha L. P. MacLeod, UNBC; Pouyan Mahboubi, Northwest Community College; Indrani Margolin, UNBC; Cathy Ulrich, Northern Health

**Chapter 6**

Philip J. Burton, UNBC; Alana J. Clason, UNBC; Stephen J. Déry, UNBC; Maya K. Gislason, UNBC; Sybille Haeussler, Skeena Forestry Consultants; Kathy J. Lewis, UNBC; Nicole M. Lindsay, UNBC; Kendra Mitchell-Foster, UNBC; Phil

M. Mullins, UNBC; Bram F. Noble, University of Saskatchewan; Katherine L. Parker, UNBC; Ian M. Picketts, Quest University; Nobuya Suzuki, UNBC; Pamela A. Wright, UNBC

The process of putting together a book project such as this always requires a great deal of assistance and advice along the way. We thank our publisher, Springer International Publishing AG, for their support. Eric Hardy, Springer Science + Business Media, provided assistance with manuscript preparation. Special thanks to Janet Slobodien, Editor for Ecology & Evolutionary Biology at Springer Science + Business Media, who was our managing editor and who had faith in our project from the start and helped us navigate the path from concept to finished book. We also thank Joanna Paterson for her careful assistance in collating, preparing, and formatting the manuscript to meet Springer specifications.

In acknowledgement of their direct and indirect contributions to the book, we also thank Jamie Reschny, Julia Russell, Esther Tong, and Homaida Razack. We have undoubtedly omitted others who deserve recognition (and to whom we apologize!) and want to emphasize that a book such as this is only possible through the collective effort and support of many people.

In finalizing the book, we asked a number of colleagues to provide a high-level review of the draft manuscript to ensure that we were reaching our target audiences. Despite extremely short timelines, we received excellent perspectives and helpful comments from Bram Noble, University of Saskatchewan, Lars Hallstrom, University of Alberta, and Marvin Eng, British Columbia Forest Practices Board. That said, they bear no fault or responsibility for any shortcomings or errors in the book, as these rest with the authors alone.

Part of our work in this book draws specific inspiration from the leadership shown by colleagues at UNBC, such as Lito Arocena, in showing how integration is indeed an imperative.

# Contents

## Part IV Synthesis

# Abbreviations

| | |
|---|---|
| BC | British Columbia |
| CEA | Cumulative effects assessment (or analysis) |
| CEAA | Canadian Environmental Assessment Agency |
| CEAMF | Cumulative effects assessment and management framework |
| EA | Environmental assessment |
| EIA | Environmental impact assessment |
| GIS | Geographic information system |
| HHRA | Human health risk assessment |
| HIA | Health impact assessment |
| IPP | Independent power project (or producer) |
| LNG | Liquefied natural gas |
| LRMP | Land and resource management planning |
| MCEC | Manitoba Clean Energy Commission |
| MKMA | Muskwa-Kechika Management Area |
| NH | Northern Health |
| NWT | Northwest Territories |
| REA | Regional environmental assessment |
| ROR | Run-of-river |
| RSEA | Regional strategic environmental assessment |
| SIA | Social impact assessment |
| TEK | Traditional ecological knowledge |
| UNBC | University of Northern British Columbia |
| USA | United States of America |
| VC | Valued component |
| VEC | Valued ecosystem component |
| WCGT | Westcoast Connector Gas Transmission project |

# List of Figures

**Fig. 2.1** A Landsat Thematic Mapper satellite image (1:100,000 scale) illustrates the cumulative effects (i.e. *nibbling loss*) of forest harvesting, road construction, and petroleum exploration and development across an area of central Alberta (Landsat imagery courtesy of the U.S. Geological Survey). These activities will have both positive (e.g. economic development) and negative (e.g. loss and fragmentation of habitat for species that depend on old-growth forests) impacts for the region ...................... 23

**Fig. 2.2** The 287 kV Northwest Transmission Line, under construction in this image, will stretch 344 km across an undeveloped but mineral-rich region of northwestern BC (Photo by C. Johnson). The transmission line will provide electricity to planned and future mines as well as an interconnection point for run-of-river hydroelectric development. The right-of-way and associated infrastructure (effects) will have direct impacts on biodiversity and other values. More complex, however, are the indirect economic growth-inducing effects resulting from the provision of inexpensive electricity to large industrial projects. The positive and negative impacts of future resource development will greatly exceed the impacts of the transmission line; however, those impacts are difficult to predict and consider during project assessment and approval ...................... 27

**Fig. 2.3** Projected development of oil and natural gas reserves in Canada (data from NEB et al. 2013) ...................... 29

**Fig. 2.4** Projected production of electricity in Canada, by source (data from NEB 2013) ...................... 29

**Fig. 2.5** Economic benefits of mineral production and the export of forest products from Canada (data from NRCAN 2013a, b) ...................... 31

**Fig. 3.1** An illustration of the types of impacts that may occur for wildlife species with increasing magnitude of the effects, with larger scale and longer-term effects, or with a combination of the two. Simple, small-scale, short-term effects may influence only the behaviour of individual animals, perhaps by displacing them from a local area. As the duration and scale of these effects increase and accumulate, the impacts on wildlife species begin to affect their condition (i.e. their physiology, energetics, and nutrition), followed by their vital rates (e.g. recruitment, survival), eventually leading to population-level and even long-term community-level impacts (after Johnson and St-Laurent 2011)........ 59

**Fig. 5.1** The *Canadian Environmental Assessment Act* (Government of Canada 2012) defines a process that has six major steps: (1) proposal and description of the project by a proponent; (2) screening by the CEAA; (3) delegation of authority for the review; (4) scoping to identify the factors to be considered; (5) expert studies and (6) decision on the project's fate. Within these steps, the *Canadian Environmental Assessment Act, 2012* identifies a series of discrete points where health considerations may be addressed (*yellow shading*), though these are not necessarily triggered in all environmental assessment processes. ERA, ecological risk assessment; HHRA, human health risk assessment; P, opportunity for *public* input where health concerns may be identified as factors to be considered in the EA. See Box 5.1 for details of these steps........ 126

**Fig. 5.2** Resource development and the cascade of effects and impacts. Upstream drivers of change in a landscape or region of concern influence policies or projects that lead to a proposed or actual change. The integrative metaphor of a cascade depicts the downstream effects and impacts arising as a consequence of the proposed or actual change, and links community and environmental impacts with the social and environmental determinants of health........ 142

**Fig. 6.1** The Nechako Basin (major lakes and rivers shown in *blue*) drains both the Nechako and Stuart-Takla watersheds. Covering approximately 52,000 km$^2$, it encompasses primarily gently rolling coniferous-forested lands along with mountainous terrain in the basin's headwaters, and flows into the Fraser River at its confluence in Prince George, BC........ 156

**Fig. 6.2** The Muskwa-Kechika Management Area (*shaded grey*) is located in northern BC. The area is comprised of a combination of provincial parks, protected areas, and special management areas (see text) and is a globally significant area of wilderness, wildlife, and cultures, to be maintained in perpetuity. Resource development and other human activities can occur in portions of the area provided that wildlife and wilderness values are maintained. Major roads are depicted throughout the province, but only a portion of the Alaska Highway intersects the Muskwa-Kechika Management Area ........................... 159

**Fig. 6.3** Cumulative environmental effects from natural and anthropogenic sources are causing rapid decline of whitebark pine populations and ecosystems. Feedbacks among predisposing (*in dark blue*), triggering (*light blue*), and accelerating factors (*in grey*; see text) may have precipitated an extinction spiral ........................... 165

**Fig. 7.1** Moving beyond the limitations of the singular gaze. The shaded arc represents a limited singular gaze, highlighting one project, within one industry, focused on one resource, one community and one kind of impact assessment and excluding things outside of this view. Cumulative thinking begins when we shift our gaze to acknowledge the wider backdrop of change, and bring other interrelated factors into view. This encourages explicit connections across projects, industries, resources, communities, and assessments (*horizontal arrows*), and also connections among these interacting factors (*vertical arrow*) ........................... 207

**Fig. 8.1** Conceptual integration of the necessary elements of an integrative regional cumulative impacts framework. Progression in the development and implementation of framework is represented from *bottom* to *top* by *ovals*. Key elements for development of the framework are presented in *rectangular boxes*, and *large green arrows* represent key elements for supporting the developed framework. Example of regionally important values could include water quality, employment, species at risk, and recreation. The *background hexagon* indicates the overriding importance of effective governance as well as the potential for the complementarity of cumulative impacts frameworks that may differ, but are adjacent across regions ........................... 225

# List of Tables

# List of Boxes

# List of Vignettes

# Part I
# Cumulative Impacts

# Chapter 1
# Cumulative Effects and Impacts: The Need for a More Inclusive, Integrative, Regional Approach

**Greg R. Halseth, Michael P. Gillingham, Chris J. Johnson, and Margot W. Parkes**

## 1.1 Introduction

The decisions we make today have implications that reach long into the future. This is especially the case for resource development projects. For example, decisions to harvest trees have immediate effects on ecosystems, economies, and communities, but the impacts of, for example, increasing access to backcountry areas will be experienced over the coming decades and it can take up to 100 years for the forest ecosystem to fully regenerate. Other resource development projects, such as mining or extraction of oil and gas, also have both immediate effects as well as impacts that range across time. These examples may seem quite obvious. Less obvious are the combined longer-term socioeconomic and ecological implications that flow from decisions to allocate timber licences or to sell mining or oil and gas leases. Such *allocation of rights* decisions effectively pre-determine the type of resource development pressures a region will experience long before any form of discussion, dialogue, or debate on the development pathway or project assessment process gets started. In these cases, the choices both now and into the future are already constrained.

Like so many key terms in common use, *effects* and *impacts* have been employed in a range of different ways in research, in debates, by different writers, by different stakeholders, or in different legislation. As such, the two terms are sometimes defined, but sometimes not; moreover, they are sometimes used discretely, but are sometimes assumed to be synonymous. Although Chap. 2 provides a more detailed exploration of these definitional challenges, in this book we define an *effect* as a direct and observable change in the current circumstance, whereas an *impact*

G.R. Halseth (✉) • M.P. Gillingham • C.J. Johnson • M.W. Parkes
The University of Northern British Columbia, Prince George, BC, Canada
e-mail: Greg.Halseth@unbc.ca; Michael.Gillingham@unbc.ca; Chris.Johnson@unbc.ca; Margot.Parkes@unbc.ca


M.P. Gillingham et al. (eds.), *The Integration Imperative*,
DOI 10.1007/978-3-319-22123-6_1

represents the longer-term consequences that flow from that change. Impacts are much wider and more nebulous, and oftentimes they are much more difficult to discern. These impacts also range across environmental, community, and health issues; more often than not, they also create linkages among these issues. The challenges they pose is why our focus in this book is on impacts.

Today, there are many forms of environmental assessment (EA) processes in use. Most follow the long-established format of focusing only on individual resource development projects and starting only after a proponent tables a development proposal. Following those early roots, these assessment processes focus almost exclusively on environmental topics, and are generally oriented toward identifying tangible effects and proposing management options for mitigating them. Even though society has changed considerably over the past 40 years, our assessment processes have changed only incrementally. Society has a much greater awareness of environmental issues and the interconnected web of impacts that development activities have on our land, water, and air over different timescales, and there is much greater awareness that these impacts extend beyond the physical and natural environment. We understand much more about how communities, families, social systems, and local cultures are affected. We understand much more about the interactions between economic and environmental health, and among ecosystem, community, and individual health. We understand much more about how widely these community and health impacts vary across both temporal and spatial scales.

While science and society have advanced our understanding of the complexity of environmental, community, and health impacts, regulatory assessment processes based strictly on the requirements of legislation generally remain our only venue for debate about these impacts. The process was generally not designed for, and is not equipped to deal with, the range of expectations now placed upon it. The process is certainly not equipped to be a forum for dialogue and learning. This mismatch is the foundation for our argument in this book that evolutionary changes in assessment processes must be replaced with revolutionary changes that will lead us towards more comprehensive, integrative, regional impact assessment. In this book, we combine our interest in environmental, community, and health impacts to help extend the conversation towards revolutionary change.

This book is also premised upon the recognition that the accumulation of impacts from multiple resource development sectors and from multiple resource development projects reduces our available future options over time. Developments may involve large or small projects, or a combination of the two. In many regions, despite the scale of current development, diverse future options and choices are still possible. The imperative, however, is to act now to conceptualise and test new approaches to understanding the impacts of natural resource development that can better support debate and decision making. In this book, we explore and define a broader and more integrative vision for identifying and then managing the regional cumulative impacts that result from natural resource development. To that more integrated regional approach, we add a longer time horizon so as to incorporate the cumulative trajectories of past, current, and future impacts.

The purpose of this chapter is to provide an introduction to the book, define the scope of the challenges and opportunities related to impact assessment in general and cumulative impact assessment in particular, and provide a roadmap to help readers address the issues these impacts raise. The current acceleration of global resource development, particularly in terms of energy exports, means that it is crucial for us to start new conversations and frame new approaches. We present ideas that are informed by the sense of urgency created by multiple, concurrent resource development trajectories in northern British Columbia (BC), Canada, but recognise that these ideas could be applied and contextualised in any region that faces similar concerns. We also present ideas, informed by a review of the literature in an even wider range of contexts. On this basis, we suggest that it has long been recognised that evolutionary change in what have become our standard toolkits for impact assessment has not been sufficient and that we instead need a revolutionary change in our approach to understanding integrated, regional, and cumulative impacts.

## 1.2 Time and Timing

In this section, we highlight several key factors that will inform our emphasis on an integrative regional approach to understanding the cumulative impacts of resource development. Time and timing are central to the challenge of rethinking our approach to impact assessment.

### *1.2.1 Why Now?*

The challenge with existing development or impact assessment processes is that they are, in general, project-based and reactive (see Vignette 6.8 in Chap. 6). That is, the evaluation of potential impacts tend to occur after a development has been proposed and after it has entered some formal regulatory process; the evaluation is not a proactive part of the original proposal. In addition, multiple projects that are being developed (or that are already in production) in the same region are generally treated as independent rather than inter-related. Perhaps most important—and least well recognised—is the fact that most assessment processes are initiated long after some of the most critical decisions have been made. These decisions, such as the allocation of mineral rights or land leases, set in motion a series of pre-determined activities that lead towards a particular type of development debate and trajectory (see Vignette 6.1).

Recently, a number of new initiatives and programmes have addressed impact assessment and the related assessment processes. For example, work has started in BC on a cumulative effects framework to better support integrated resource management and decision making for the natural resource sector (Government of BC 2014). In Ontario, the provincial government created a Ring of Fire Secretariat in 2011 within its Ministry of Northern Development and Mines to help connect development processes with regional communities and their concerns about (and processes for) environmental impact assessment (EIA; Government of Ontario

2015). In Manitoba, a recently initiated assessment process for the cumulative impacts of hydroelectric developments within the Churchill, Burntwood, and Nelson river systems has adopted both a regional and a historically informed approach (Manitoba Hydro 2014). In Alberta, the new South Athabasca Oil Sands (SAOS) Regional Strategic Assessment and Sub-regional Plan process focusses on the potential cumulative impacts "of three energy development scenarios … [and] is intended to increase understanding of the potential social, economic and environmental effects of in situ oil sands activities in the SAOS area before they occur" (Government of Alberta 2014, p. 2). In Australia's Moranbah region, there were challenges convincing industries that even if individual mines had not exceeded legislated pollution levels, the cumulative impact of dust and pollution had exceeded acceptable health levels. The Moranbah Cumulative Impacts Group (2015) has been working to bring industry stakeholders together to investigate noise, dust, and air quality issues.

In our view, this global interest, and the degree of experimentation (e.g. spatial, temporal, proactive planning) within such initiatives, reinforces our argument for the value and timeliness of this book and for more critical reflection on impact assessment issues that can point towards a more comprehensive understanding, better decision making, and better management. We need to find a way to become more proactive before too many initial steps cast particular development directions in stone and thereby limit our future alternatives.

As detailed in later chapters, the current array of environmental, community, and health impact assessment processes have each had a relatively recent genesis; that is, they have developed out of a particular set of problems that previous processes could not manage. Of particular interest in this book is the Berger Inquiry into the Mackenzie Valley Pipeline proposal in northern Canada (Berger 1977). Conducted over more than 3 years in the mid-1970s, the inquiry resulted in the two-volume *Northern Frontier, Northern Homeland* report that not only detailed matters of concern around the environmental, community, and health impacts of that project, but also outlined a new way of assessing and evaluating impacts. Most current Canadian environmental, community, and health impact assessment processes were inspired by the Berger Inquiry and have now been codified into legislation and the resulting regulatory processes (see Chap. 7). We argue that the well-described problems with current assessment processes have accumulated to the point where tinkering is no longer viable. What is needed is a *neo-Berger* revolution that involves a complete rethinking of how we undertake impact assessments. We recommend a more integrative, regional, and strategic approach for assessing, planning, and managing the cumulative impacts of natural resource development.

### 1.2.2 Past–Present–Future

In considering the impacts of resource development, we cannot be ahistorical. Instead, we are challenged to find approaches and processes that explicitly consider the past, the present, and the future at the same time. This means that decisions in the present need to be simultaneously informed by both the decisions (and impacts) of the past, and the implications of these present decisions on options for and in the future.

The time-lag inherent in the differences between effects and impacts also cannot be ignored. Short-, medium-, and long-term implications need to be considered as we look ahead and consider environmental, community, and health impacts. Recognition of such timing and time-lag issues requires assessing and managing impacts at a range of timescales that may range from hours, days, and weeks, through to months, years, and decades; for some impacts, the appropriate scale may be centuries.

Attention to timescales becomes more complex when considering the impacts of both current and past decisions and activities. At any one moment, and in any one context, a combination of short-term impacts from one decision, medium-term impacts from another decision, and long-term impacts from a different decision may converge to influence what is a reasonable decision today. Blindness to this time-lag, and to how impacts unfold across different timescales, can have serious consequences.

### 1.2.3 Time, Pressure, and Pace

This book also responds to another important temporal consideration in relation to the perceived *pace* of decision making. For many who are observing and interacting with current environmental, social, or health impact assessment processes, there is a sense that decision making is either *speeding up* or is under pressure to speed up.

Such perceptions of a more rapid pace may be the result of several factors (see Box 4.3). One may be that our increased access to information means we are more aware of these processes and the associated deliberations, as well as how new knowledge or research developments may or may not be informing these processes and deliberations. Another may be that there are generally many public processes simultaneously underway across a range of environmental, community, or health issues. Few of these governmental, institutional, or community-driven processes have the time or capacity to take stock of the interactions among these discrete time-pressured processes. People are concerned about having the time to listen, learn, understand, discuss, and then contribute meaningfully when the demands of involvement seem to be growing at an accelerating pace. When people lack the time, resources, or mandate to engage with the links between environmental, social, and health concerns, they often ignore the reality of complex and interrelated issues (see Chap. 7). Another source of the feelings of acceleration may be the repeated calls in recent years for more streamlined processes by some governments and project proponents. The saying that *time is money* can seem very much the case when resource development projects are linked to short-term windows of opportunity in global markets. These calls for streamlining are often associated with concomitant pressures towards an efficient process that may be achieved at the cost of effectiveness, especially in relation to the need for time to gain a more comprehensive understanding of the cumulative implications of resource development.

### *1.2.4 Timing of Involvement*

Accompanying the challenge of time pressure and the accelerated pace of decision making is the question of who should be involved, and at what point in the process. This book is premised on the fact that the disconnect between the technical process of EA and the wider decision-making processes can limit opportunities for the most relevant actors (and knowledge-holders) to be engaged and involved at the right times. This challenge is integrally related to the larger question of where impact assessment fits within broader processes of creating a vision to guide development and the governance of that development, and the need to reconsider the timing of both. Questions around the nature and timing of involvement reflect growing attention to questions related to *governance*, which represents the processes whereby societies or organisations make their important decisions, determine whom they should involve in those processes, and how they render account (Graham et al. 2003; Jordan et al. 2003; Howlett et al. 2006).

Given the difficulties of making appropriate decisions within the time constraints of economies and societies, we also challenge the attitude that we should not bother because we are already too late. Given the time-lags inherent to some impacts, and the commitment to not further reduce our future options by making short-sighted or rushed decisions, there is always room for a more comprehensive understanding. The dialogue to change the way we do things needs to start somewhere—so why not *here* and why not *now*? If we were to look back from 50 years in the future, how would we judge our failure to take this opportunity to explore and test more integrative regional cumulative impact assessments?

## 1.3 An Integrative Focus

Impact assessment today is stuck. As detailed in the following chapters, it is limited by its focus on individual large development projects and its failure to address a suite of spatial and temporal scales as well as the full set of environmental, social, and health values. Many are working to improve current environmental, social, and health assessment processes, but they are almost all working within the existing *box*; they are tinkering with an approach that is increasingly acknowledged to be broken. Throughout this book, we have tried to bring something new to the debate about the efficiency and effectiveness of current impact assessment processes. Our focus is on integrating environmental, community, and health perspectives into a broader, more inclusive, and holistic understanding of cumulative impacts.

Specifically, we are proposing an integrative approach to understanding and then addressing the environmental, community, and health impacts of natural resource development.

Our use of the word *integrative* is an intentional, active choice that accounts for the complexity of environmental, community, and health systems and processes. We are aware that integration is an ongoing (and societal) process that is rarely achieved in a way that the word *integrated* may suggest. But it is vitally necessary to bring together the imperatives and processes of these three interacting systems if we are to trace the impacts of our choices and actions in a more inclusive and comprehensive way.

Although we are proposing something new, it is important to emphasise that this breadth of interest runs parallel to other transformative debates, including the *triple bottom line* or the *three pillars of sustainability* that each link to economic, social, and environmental issues, and growing attention to resilience and health within social–ecological systems. At a more general level, our proposal recognises the economy as a significant driver of change, but is informed by growing interest in broadening the understanding of economic development options and pathways by considering the future in relation to a wider range of environmental, community, and health issues and perspectives. The integrative challenge we lay out in this book can, therefore, be seen in parallel to long-standing debates and concerns, but focused intentionally on the specific challenge of integrating environmental, community, and health dynamics.

## 1.4 Vision and Scale

Debates and conflicts over natural resource development often come down to different visions about how the natural assets of a locality or region might be mobilised to meet the aspirations of different interested parties. These visions are reflections of values, information, experience, pressures, concerns, and past debates. A critical challenge that will emerge over time is that as societies and their economies transform, and as the underlying values of society change, we need to adjust and fine-tune our sense of what we wish to achieve with the assets and within the constraints of our socioecological landscape.

There are historical examples of development policy approaches that were well-tuned to the economies and societies of their day. From the close of the Second World War until the 1970s, for example, the industrial resource development model employed in BC was tuned to the expanding industrial output of developed nations, the resource-supply warehouse opportunities of the region, its proximity to manufacturing centres in the United States of America (USA) and Asia, the need for expanded employment and revenue to support the province's baby-boom families, and the appetite of the province's population for a greater range of public services and investments to support an improved quality of life (BC Post-War Rehabilitation Council 1943; Mitchell 1983; Tomblin 1990; Williston and Keller 1997). Since the

1980s, however, resource-producing regions such as BC and areas of Australia and the USA have not undertaken a purposeful exploration of how such a vision might fit with the contemporary directions of global economies or the contemporary expectations of our communities and societies.

Without a vision, there is no foundation by which to evaluate competing or converging interests, there is no way to identify and describe the values that we consider important, and no way by which one might measure progress towards fulfilling that vision or protecting those values. There is also no framework for evaluating the trade-offs to be made between who shares in the benefits and who bears the costs, or for evaluating how these benefits and costs are spatially and temporally distributed (see Vignette 6.3). In a setting with competing values and visions, it is crucial to get these considerations onto the table as part of the process of building understanding and moving towards greater consensus. In addition, we are challenged with the reality that visions related to environmental, community, or health topics are each sensitive to different scales. Visions enacted at a state or provincial scale may or may not align with the vision of smaller regions or of communities where the impacts of resource development are experienced *on the ground*. Similarly, visions at a national level face the same potential for connection or disconnection across these scales. We argue that part of the skill set needed to develop a vision appropriate for the twenty-first century is the ability and aptitude to *zoom in* and *zoom out* across spatial and temporal scales, thereby acknowledging the range of perspectives this demands, and the important cross-scale interactions that are now a feature of governance processes from local to global.

## 1.5 Development Choices: Jurisdiction, Priorities, and Responsibilities

In the absence of a wider (state or provincial rather than local or regional) vision, development choices may or may not serve the long-term interests of the landscapes and communities directly affected by the choices. Without a vision to lend coherence and consistency to the approach, decision making is more likely to be influenced by the disjointed array of jurisdictional mandates and their self-generated needs, priorities, and responsibilities. It is also more likely influenced by short-term crises or needs without consideration for what might be the better course over the long term. Particularly problematic is that many of these self-generated needs or priorities have been shown to be very short-term and even transitory. Such priorities fly in the face of what is known about the long-term and interwoven pathways by which effects become environmental, community, and health impacts. As we and our colleagues highlight throughout this book, lessons from the past show that we are more successful and better able to integrate environmental, community, and health concerns and opportunities when we adopt a proactive approach based on a more comprehensive vision and understanding (see Vignettes 6.1 and 6.2).

A further challenge that arises from jurisdictional fragmentation is that the mobilisation of different natural resource development projects now occurs under different frameworks. Many processes and governments are struggling with how to coordinate resource development across activities and sectors. There is a crucial connection between mobilising visions and having consistent development frameworks that allow an assessment of conformity with that vision. Instead, we are faced with different choices being made about different resource development options (often across different resource development sectors), based on different frameworks, jurisdictions, and priorities, each with different responsibilities and capacities for assessing and monitoring.

A second critical issue related to development alternatives involves the conflicting roles of national governments, which are becoming more apparent globally as converging resource development highlights the multiple roles and responsibilities of government actors. The state may be simultaneously a regulator of development, the beneficiary of that development, the agent that is actively moving the development process forward, the arbiter of the review or assessment process that provides only an *opinion* to government, and the agent responsible for monitoring compliance and impacts.

### 1.5.1 Jurisdiction, Rights, and Title

The degree to which the development visions of states or industries fit with local aspirations, and the degree to which local and regional voices are able to participate in the dialogue about impacts vary by project, by type of natural resource activity, and by the complexity of the jurisdictional landscape. An additional complexity for integrative regional cumulative impact assessment derives from the critical jurisdictional issue around the involvement of Aboriginal peoples (see Vignette 6.6; see also Boxes 3.6, 4.3, and 5.4). In many Organisation for Economic Cooperation and Development states, these are special considerations. In northern Norway, for example, the Sami Parliament, the Finnmark Estate Agency, and the Finnmark Commission are examples of additional jurisdictional bodies that have a role to play in evaluating natural resource development (Lund 2015; Mikaelson 2015; Pedersen 2015). In Australia, new mining and natural gas developers often negotiate local or regional benefits agreements under the auspices of the federal *Native Title Act* (Government of Australia 1993). Although these agreements have become especially common in the booming resource development regions of Queensland and Western Australia, there continues to be contention, debate, and legal action around implementation of the development plans (Trigger et al. 2014).

In much of Canada, treaties between Aboriginal nations and the government were signed before intense European colonisation and settlement began. In BC, however, such treaties were not universally signed. Successive provincial governments historically used the idea of *empty lands* to argue that Aboriginal title did not exist in the province and to justify the social and economic marginalisation of Aboriginal peoples. The empty lands thesis "legitimized the denial of Aboriginal title and sanctified the new white doctrine that all land in the colony was not only under British sovereignty but also directly owned by the Crown" (Tennant 1990, p. 41; see also Harris 1997, 2002). One result is that debate and legal action over rights and title to the land continue. Court decisions have gradually recognised, affirmed, and defined Aboriginal title, thus requiring the current consultation and accommodation processes related to major resource development projects in both treaty and unceded territories (First Nations Summit 2005; BC Government 2007; BC Treaty Commission 2014). Not only does this complicate natural resource development assessment processes, but it politicises them as well.

## 1.6 Assessment Processes: Effects and Impacts over Time and Space

The challenge of overlapping jurisdictions, priorities, and responsibilities highlights the importance of understanding how different levels and scales of resource management decisions affect the cumulative environmental, community, and health impacts. As detailed in later chapters, there exists a significant body of work on the existing assessment processes, and on the definitional issues associated with the concepts of effects and impacts (including how they become cumulative). We detail the historical development of these processes, together with the common and converging critiques made of current approaches, as a foundation for our proposal to establish a more integrative approach to understanding regional cumulative impacts.

Furthermore, the following chapters highlight the possibilities for amending the existing processes and approaches. Throughout these chapters, two recurring themes include: the need for explicit consideration of the spatial and temporal contexts of development decisions (i.e. taking into account past, present, and future implications of institutional structures and their decisions within and among particular places and regions) and the need for assessment processes that integrate a consideration of environmental, community, and health impacts.

Our focus upon the environmental, community, and health perspectives promotes the notion of undertaking a more integrative and recursive approach to developing a fuller understanding of the impacts. This also leads to the recognition that no matter how comprehensive or thorough any assessment is, it will never be sufficient without a commitment to learning and adaptation (including explicit recognition that existing structures and processes will eventually need to be refined and revised). Moreover, it will not be sufficient if it is considered in isolation from the broader institutional and governance processes within which it is embedded. Not only does our approach need to recognise cumulative impacts, but we also need to apply *cumulative thinking* to how to conceptualise both process and approach (see Chap. 8).

## 1.7 Complex Concepts

Before we describe the outline of the book, it is important to recognise that throughout its contents we wrestle with complex concepts. This complexity can lead to confusion, especially if different people have different understandings of the same words and terms. As detailed throughout the book, existing critiques of environmental, social, and health impact assessments suggest that tinkering with existing frameworks is not sufficient in the face of long-standing challenges. In response, our proposal is to replace existing approaches with a more integrative regional cumulative impact assessment. As readers will no doubt recognise, each of the terms *integrative*, *regional*, *cumulative*, and *impact* are complex. Although these terms are unpacked in subsequent chapters, and especially Chaps. 2–5, we summarise these points here in Box 1.1 to provide an overview.

**Box 1.1 Our Key Terms: Integrative, Regional, Cumulative, and Impact (Greg R. Halseth)**

**Integrative:** We chose the term integrative to remind us of the need to look broadly across environmental (see Chap. 3), community (see Chap. 4), and health (see Chap. 5) impacts. This is important because we recognise that these impacts may be quite different over time, and may be differentially felt, in the environmental, community, and health spheres. Also, impacts specific to each sphere will interact, leading to even more complex, but real, outcomes for environments and communities.

**Regional**: We recognise that the economy and the economic implications of decisions are key drivers of change in a global capitalist economy. We also recognise through a place-based framework that places are the locales in which the decisions, policies, and imperatives of regional, provincial, national, and international activities play out. Places, however, suggest a relatively discrete area, and are therefore too small of a scale for an integrative cumulative impacts framework. Instead, we argue for a larger *regional* approach to understanding more fully the consequences of the historical trajectories of, and the future impacts from, natural resource development.

**Cumulative**: We support the notion of *cumulative* impacts by recognising the importance of impacts that result from both single resource development activities and from those that are cumulative across projects and development sectors at a range of spatial and temporal scales. We also support the need to explicitly include environment, community, and health impacts. Furthermore, we note that even many small and relatively unobserved effects can generate cumulatively significant impacts over time. We recognise the critiques of existing assessment processes and the associated calls for a wider, deeper, and longer-term perspective.

**Impacts**: As noted above and further defined in Chap. 2, an *effect* is a change to the current circumstance—something that shows direct and observable

(continued)

**Box 1.1** (continued)

changes from resource development. In contrast, an *impact* represents the consequences of the effect and is seen through the lens of a value or an entity affected by that change in current circumstances. As such, these impacts are multiple, interwoven, and complex, changeable over time, and need to be traced through local to (at least) regional scales. Impacts can be positive or negative, depending on the set of values in play and the perspectives of those who are involved in a decision-making process. They are wider and sometimes much more difficult to discern than effects. It is for these reasons that we focus on impacts in this book.

Other terms are central to our argument. Nested within our regional approach is the notion of *community*. We take an inclusive approach to the notion of community, one that involves both place-based and interest-based definitions (see Chap. 4). We also explicitly recognise that community must include both Aboriginal and non-Aboriginal communities.

In terms of Aboriginal communities, we recognise that different jurisdictions will define Aboriginal peoples differently under their respective constitutions. In many colonial contexts around the world, the terminology that describes Aboriginal and colonial (settler) peoples is controversial, contested, and changing (Harris 2002). The situation is made more complex by layers of history and accumulated legal structures. As a result, there is no one terminology that is accepted or applied globally, and no one approach that is accepted or applied to understanding this terminology. In this book, we use the terms Aboriginal, First Nations, and Indigenous in different contexts. In Canada, much of the conversation over terminology is guided by the Canadian Constitution (Government of Canada 1982). Part II, section 35, of the *Constitution Act* of 1982 defines the Aboriginal peoples of Canada to include the Indian, Inuit, and Métis peoples of Canada:

> "Aboriginal peoples" is a collective name for the original peoples of North America and their descendants. The Canadian constitution recognizes three groups of Aboriginal people: Indians (commonly referred to as First Nations), Métis and Inuit. These are three distinct peoples with unique histories, languages, cultural practices and spiritual beliefs. (http://www.psc-cfp.gc.ca/plcy-pltq/eead-eeed/dg-gd/aaa-bg-dr-eng.htm)

As this quote shows, the term First Nations is in common use in Canada, but it has very specific political connotations related to the original nations before European contact, and is often used in matters of government authority and legal jurisdiction. In northern BC, 17.6 % of the population identifies as Aboriginal (BC Stats 2006). As described by Willems-Braun (1997, p. 26):

> First Nations is the term preferred by BC Aboriginal communities, deliberately subverting the primitivist tropes of "tribe" found in anthropological literatures. In addition, the term is used to assert an organized, political presence that preexists European contact while simultaneously placing in question the territorial claims of the Canadian nation-state.

At an international level, the term *Indigenous* is more commonly used as an inclusive term. This is especially the case since the 13 September 2007 *United Nations Declaration on the Rights of Indigenous Peoples* (United Nations 2007). Throughout this book, we primarily use the term Aboriginal, recognising that in the Canadian context, this term includes First Nation, Métis, and Inuit people, and is generally considered as the Canadian synonym for Indigenous. Depending on context, however, we also use First Nation (Canadian political contexts) and Indigenous (international contexts).

The next group of terms includes government and governance (Jordan et al. 2005). These terms define quite different things (Jessop 2001; Brenner et al. 2003). *Government* refers to the legally constructed and spatially constrained jurisdictional bodies that administer various rights and laws across a designated territory. In most national settings, there is either a two-tier (national versus local government) or a three-tier (national versus state or provincial versus local government) administrative hierarchy (Tindal and Nobes-Tindal 2004; Douglas 2005; Shucksmith 2010). Most current environmental, social, and health impact assessment processes are administered by national, or state/provincial levels of government.

The concept of *governance*, however, is different. At its simplest, it refers to the process of managing participation and involvement, leading towards decision making (Rhodes 1996; Marsden and Murdoch 1998; O'Toole and Burdess 2004). The concept of governance is increasingly important in resource development and in processes focused on assessing the regional cumulative impacts of such development (Wilson 2004; Parkins 2008; Cheshire 2010). A wider range of groups and interests than in the past now wish to participate in debates, dialogues, assessments, and other related processes for governing and monitoring resource development (MacKinnon 2002; Bryant 2011). The inclusion of these new voices is part of a long-term policy shift, but it means that our previous focus on singular government actors must now give way to a broader group of actors and mechanisms of governing resource development. Echoing our calls for revolutionary change in undertaking cumulative impact assessment, Bartlett (1993, p. 162) argued from a policy concern with environmental problems and a governance concern with integrated impact assessment because "incremental innovation is seldom able to affect significantly the tendencies of the larger system of public policy and administration within which it is undertaken".

## 1.8 Outline of the Book

This book is structured into four parts. The first part includes the present chapter and Chap. 2, which provides critical background and context for the issue of cumulative impact assessment. Chapter 2 opens with the recognition that cumulative change is not a new concept. In Canada, for example, we have extensive experience with environmental, social, and health assessment processes that consider cumulative impacts. Despite such experience, there is much criticism of existing processes and repeated calls for improvements or changes. To establish a framework for our exploration of the needed changes, Chap. 2 sets out the fundamental concepts underlying cumulative effects and impacts, as well as how they have been codified into legislative and regulatory frameworks.

The second part of the book brings together three very different perspectives for understanding cumulative impacts. The environmental perspective (Chap. 3) is the one most often linked to impact assessment work, but is also one that continues to identify critical shortcomings in current practices. Although the cumulative impacts of natural resource development have been well-studied, current assessment and approval processes continue to operate on a project-by-project basis with highly circumscribed geographic and temporal parameters. We need to better understand the cumulative scope, scale, and legacy of impacts over time. Chapter 4 introduces a community perspective. Not only does this perspective provide both a spatial and a conceptual *place* for understanding how impacts accumulate over time, it also describes how they intersect with community development. Chapter 5 introduces a health perspective. It provides an orientation to contemporary debates about cumulative health impacts as an integrative challenge, and introduces the idea of these health impacts as being linked to environmental and community concerns through a cascade of impacts on the social and environmental determinants of health. Chapters 3–5 include short contributions in the form of *Boxes* that illustrate the key challenges underlying current processes for identifying, understanding, or managing cumulative impacts. Our critique of current frameworks and approaches draws from this unique collaboration among multiple perspectives.

Part III of the book provides a more practical, and sometimes personal, "on the ground" orientation. A series of eight *Vignettes* (Chap. 6) are used to provide examples of cumulative impacts that illustrate a number of recurrent messages about the changes needed to provide a more integrative and effective assessment process. Some of the most dramatic cumulative impacts are associated with waterways and watersheds. The opening Vignette, *Exploring the Cumulative Impacts of Climate Change and Resource Development in the Nechako Basin* (Picketts and Déry), examines the long history of changes in the Nechako basin of northern BC against the present backdrop of climate change. One of the best ways to maintain a range of values for an area is to identify and manage or protect those values before development takes place. The second Vignette, *Maintaining Wildlife and Wilderness in the Muskwa-Kechika Management Area* (Parker and Suzuki), shares the example of one remote area where provincial law states that any industrial development must maintain wildlife and wilderness values. Cumulative impacts tied to resource-based industries are, in turn, linked to economics, and the third Vignette, *Cumulative Effects and Impacts: Influence of the Resource-Based Business Model* (Lewis), uses forest health examples to illustrate that a consideration of cumulative impacts must reach beyond direct industrial activity.

Moving beyond an exclusive focus on economic values, the fourth Vignette, *Combating the Decline of Whitebark Pine Ecosystems across Central British Columbia* (Haeussler, Burton, and Clason), examines a non-commercial tree species and its associated ecosystem, which typically fall outside of assessments, but which are very susceptible to the cumulative impacts of land activities and climate change. The fifth Vignette, *Valuing Outdoor Recreation in Living Landscapes as a means of Connecting Healthy Environments with Communities* (Mullins and Wright), explores the value of integrating nature into our daily lives through personal and societal choices, another frequently overlooked facet of cumulative impact assessment. Adding depth, the sixth Vignette, *Cumulative Environmental, Community and Health Impacts of Multiple Resource Developments in Northern British Columbia: Focus on First Nations* (Lindsay), examines why a consideration of cumulative environmental, community, and health impacts from resource development must begin with an understanding of the significance of the land to the cultural and identity dynamics of Aboriginal communities, and how these impacts are integral to individual and community health and well-being. The seventh Vignette, *Lived Reality and Local Relevance: Complexity and Immediacy of Experienced Cumulative Long-term Impacts* (Mitchell-Foster and Gislason), adds that a broader vision and scope is needed for cumulative impact analyses in order to meet the combined needs of communities, institutions, and natural systems. The final Vignette, *Scoping Out Potentially Significant Impacts: Constraints of Current Regulatory-Based Cumulative Effects Assessment* (Noble), uses two Canadian examples to illustrate the challenges of current assessment processes and how a failure to forecast impacts into the future has implications for landscapes, communities, and our health and well-being.

The fourth part of the book includes two chapters that set out our proposal for a more integrative framework that will improve our understanding of and ability to cope with regional cumulative impacts. Chapter 7 starts with our response to the challenge of enhancing an understanding of the integrative and regional dynamics of cumulative impacts. It links back to the formative years of impact assessment, and notes that there have been considerable social and policy changes that undercut the foundations of the original assessment frameworks. It also highlights the key factors driving the need to transform existing approaches and situates the notion of regional cumulative impacts as *wicked problems* that demand a new and integrative approach to building understanding.

Chapter 8, which concludes the book, opens with recognition of the need for a more integrative approach built upon cumulative thinking, and builds on past criticisms and recommendations to describe a paradigm change that is intended to transform our thinking and our approaches towards integrated regional cumulative impact assessments. The points of reference for this revolutionary change include six principles and five elements.

With cumulative thinking as our starting point, we propose six principles that should guide the development and application of region-specific frameworks:

1. We can only solve this problem through a *revolution* in strategy, not an *evolution* of practice.
2. Cumulative impacts are context-specific and are particular to regional circumstances.
3. There is an urgency to taking action to develop a vision and a framework for addressing cumulative impacts, and this urgency is exacerbated by the accelerating pace of change in the global economy—especially global resource commodity markets.

4. Only by developing inclusive processes can we identify and account for the diversity of values that are so critical to supporting sustainability.
5. Because cumulative impacts describe a *wicked problem*, our approaches to understanding and managing the problem must be iterative and involve ongoing learning, flexibility, and adaptation.
6. The processes must be transparent, as inclusiveness can only be realised if the governance structures allow for the full involvement of all representative voices.

Given these principles, there is no *one-size fits all* model for managing cumulative impacts. To support the six principles, we propose five elements that are necessary components of an effective cumulative impacts framework. These elements include:

1. Access to timely and appropriate information and knowledge;
2. The identification and measurement of agreed-upon values;
3. A commitment to monitoring and assessment;
4. The use of strategic planning and decision-making frameworks; and
5. The development of transparent and inclusive governance regimes.

Together, these five elements can be linked to each of the six principles, and the six principles, in turn, support cumulative thinking.

The book concludes with information about the contributors, and an index.

## References


Bartlett, R.V. 1993. Integrated impact assessment as environmental policy: The New Zealand experiment. *Policy Studies Review* 12: 162–176.

BC Government. 2007. New relationship. Victoria: Government of BC. http://www.newrelationship.gov.bc.ca/shared/downloads/new_relationship.pdf. Accessed 10 May 2010.

BC Post-War Rehabilitation Council. 1943. *Interim report of the post-war rehabilitation council*. Victoria: The Council.

BC Stats. 2006. Aboriginal profiles of British Columbia [Data file]. Available from http://www.bcstats.gov.bc.ca/StatisticsBySubject/AboriginalPeoples/CensusProfiles/2006Census.aspx. Accessed 11 Jan 2015.

BC Treaty Commission. 2014. *BC Treaty Commission: Annual Report 2013*. Vancouver: BC Treaty Commission. http://www.bctreaty.net/. Accessed 2 Apr 2014.

Berger, T.R. 1977. *Northern frontier, northern homeland: The report of the Mackenzie Valley Pipeline Inquiry - Berger Report*, 1st ed. Ottawa: Minister of Supply and Services Canada.

Brenner, N., B. Jessop, M. Jones, and G. MacLeod (eds.). 2003. *State/Space: A reader*. Oxford and Boston, MA: Blackwell.

Bryant, C. 2011. Co-constructing rural communities in the 21st Century: Challenges for central governments and the research community in working effectively with local and regional actors. In *The next rural economies: Constructing rural place in a global economy*, ed. G. Halseth, S. Markey, and D. Bruce, 142–154. Oxfordshire, UK: CABI International.

Cheshire, L. 2010. A corporate responsibility? The constitution of fly-in, fly-out mining companies as governance partners in remote, mine-affected localities. *Journal of Rural Studies* 26: 12–20.

Douglas, D. 2005. The restructuring of local government in rural regions: A rural development perspective. *Journal of Rural Studies* 21: 231–246.

First Nations Summit. 2005. Backgrounder – "New Relationship": Implementation of Supreme Court of Canada Decisions. http://www.fns.bc.ca/info/newrelationship.htm. Accessed 20 Sep 2005.

Government of Alberta. 2014. South Athabasca Oil Sands Regional Strategic Assessment and sub-regional plan: Fact sheet. http://esrd.alberta.ca/focus/cumulative-effects/cumulative-effects-management/management-frameworks/documents/LARP-FactSheet-Strategies-Feb13-2014.pdf. Accessed 16 Jan 2015.

Government of Australia. 1993. Native Title Act, Act No. 110 of 1993. Canberra: Government of Australia.

Government of BC. 2014. CEF Overview Report February 2014: Addressing cumulative effects in natural resource decision-making, a framework for success. Ministry of Environment and Ministry of Forests, Lands and Natural Resource Operations. http://www2.gov.bc.ca/gov/DownloadAsset?assetId=F2A8B8AE894348DBA4CF7942EC592762&filename=overview_report_addressing_cumulative_effects.pdf. Accessed 16 Jan 2015.

Government of Canada. 1982. *Constitution act*. Ottawa: Government of Canada.

Government of Ontario. 2015. Ring of Fire Secretariat. Ministry of Northern Development and Mines. http://www.mndm.gov.on.ca/en/ring-fire-secretariat. Accessed 16 Jan 2015.

Graham, J., B. Amos, and T. Plumptre. 2003. *Principles for good governance in the 21st century. Policy Brief No.15 – August 2003*. Ottawa: Institute on Governance. http://iog.ca/wp-content/uploads/2012/12/2003_August_policybrief15.pdf. Accessed 21 Nov 2014.

Harris, R.C. 1997. *The resettlement of British Columbia: Essays on colonialism and geographical change*. Vancouver: UBC Press.

———. 2002. *Making Native space: Colonialism, resistance, and reserves in British Columbia*. Vancouver, BC: UBC Press.

Howlett, M., J. Rayner, and C. Tollefson. 2006. From government to governance in forest planning? Lessons from the case of the British Columbia Great Bear Rainforest initiative. *Forest Policy and Economics* 11: 383–391.

Jessop, M. 2001. The rise of the region state in economic governance: 'Partnerships for prosperity' or new scales of state power? *Environment and Planning A* 33: 1185–1211.

Jordan, A., R. Wurzel, and A. Zito. 2003. New instruments of environmental governance. *Environmental Politics* 12(3): 1–24.

Jordan, A., R.K.W. Wurzel, and A. Zito. 2005. The rise of 'new' policy instruments in comparative perspective: Has governance eclipsed government? *Political Studies* 53: 477–496.

Lund, S. 2015. The nature of Finnmark between traditional use, international capital, and central political power. In *Sustainable Development in the Circumpolar North – From Tana, Norway to Oktemtsy, Yakutia, Russia: The Gargia Conferences for Local and Regional Development (2004–14)*, eds. T. Gjertsen and G. Halseth, 177–191. Prince George, BC: UNBC Community Development Institute Publications Series. Tromso, Norway: Septentrio Academic Publishing of the University Library at UiT The Arctic University of Norway, Septentrio Conference Series, Number 1, 2015.

MacKinnon, D. 2002. Rural governance and local involvement: Assessing state – community relations in the Scottish Highlands. *Journal of Rural Studies* 18: 307–324.

Manitoba Hydro. 2014. Regional cumulative effects assessment for hydroelectric developments on the Churchill, Burntwood and Nelson River systems: Phase 1 report. http://www.gov.mb.ca/conservation/eal/registries/5714hydro/rcea_phase1.pdf. Accessed 16 Jan 2015.

Marsden, T., and J. Murdoch. 1998. Editorial: The shifting nature of rural governance and community participation. *Journal of Rural Studies* 14: 1–4.

Mikaelson, S. 2015. Development challenges in Indigenous communities in Swedish Sapmi. In *Sustainable Development in the Circumpolar North – From Tana, Norway to Oktemtsy, Yakutia, Russia: The Gargia Conferences for Local and Regional Development (2004–14)*, eds. T. Gjertsen and G. Halseth, 216–225. Prince George, BC: UNBC Community Development Institute Publications Series. Tromso, Norway: Septentrio Academic Publishing of the University Library at UiT The Arctic University of Norway, Septentrio Conference Series, Number 1, 2015.

Mitchell, D. 1983. *WAC Bennett and the rise of British Columbia*. Vancouver: Douglas and MacIntyre.

Moranbah Cumulative Impacts Group. 2015. What is the Moranbah Cumulative Impact Group? http://mcig.org.au. Accessed 16 Jan 2015.

O'Toole, K., and N. Burdess. 2004. New community governance in small rural towns: The Australian experience. *Journal of Rural Studies* 20: 433–443.

Parkins, J. 2008. The metagovernance of climate change: Institutional adaptation to the mountain pine beetle epidemic in British Columbia. *Journal of Rural and Community Development* 3(2): 7–26.

Pedersen, S. 2015. The right to traditional resources and development programs. In *Sustainable Development in the Circumpolar North – From Tana, Norway to Oktemtsy, Yakutia, Russia: The Gargia Conferences for Local and Regional Development (2004–14),* eds. T. Gjertsen and G. Halseth, 192-200. Prince George, BC: UNBC Community Development Institute Publications Series. Tromso, Norway: Septentrio Academic Publishing of the University Library at UiT The Arctic University of Norway, Septentrio Conference Series, Number 1, 2015.

Rhodes, R.A.W. 1996. The new governance: Governing without government. *Political Studies* 44: 652–667.

Shucksmith, M 2010. Disintegrated rural development? Neo-endogenous rural development, planning and place-shaping in diffused power contexts. *Sociologia Ruralis* 50: 1–14.

Tennant, P. 1990. *Aboriginal peoples and politics: The Indian land question in British Columbia, 1849–1989*. Vancouver: UBC Press.

Tindal, R., and S. Nobes-Tindal. 2004. *Local government in Canada*, 6th ed. Scarborough, ON: Nelson/Thomson Canada.

Tomblin, S. 1990. W.A.C. Bennett and province-building in British Columbia. *BC Studies* 85: 45–61.

Trigger, D., J. Keenan, K. de Rijke, and W. Rifkin. 2014. Aboriginal engagement and agreement-making with a rapidly developing resource industry: Coal seam gas development in Australia. *The Extractive Industries and Society* 1: 176–188.

United Nations. 2007. United Nations declaration on the rights of Indigenous peoples. New York: United Nations General Assembly, Sixty-first Session, September 13.

Willems-Braun, B. 1997. Buried epistemologies: The politics of nature in (post)colonial British Columbia. *Annals of the Association of American Geographers* 87: 3–31.

Williston, E., and B. Keller. 1997. *Forests, power, and policy: The legacy of Ray Williston*. Prince George: Caitlin Press.

Wilson, G. 2004. The Australian *Landcare* movement: Towards 'post-productivist' rural governance? *Journal of Rural Studies* 20: 461–484.

# Chapter 2
# Defining and Identifying Cumulative Environmental, Health, and Community Impacts

**Chris J. Johnson**

## 2.1 Introduction

Cumulative change to the environment is not a new or unfamiliar concept. From governments to concerned citizens, there is a broad appreciation of the singular and cumulative impacts of economic development. These impacts influence the intangible and abstract elements of the natural world such as biodiversity and ecosystem health as well as those components of the environment that directly affect human health and the ability of communities to meet the socioeconomic needs and aspirations of their citizens. Indeed, cumulative impacts are recognised internationally as an outcome of development that must be planned for, regulated, and, when necessary, mitigated (Dixon and Montz 1995; Samarakoon and Rowan 2008; Zhu and Ru 2008; Retief et al. 2009; Sinclair et al. 2009; Wärnbäck and Hilding-Rydevik 2009; Kinnear et al. 2013).

In Canada, formal acknowledgement of cumulative impacts can be traced back more than three decades to the initial policy work of the Canadian Environmental Assessment Research Council (Peterson et al. 1987). In 1995, the importance of cumulative effects was recognised and entrenched in legislation when the *Canadian Environmental Assessment Act* (Government of Canada 2012) came into force and required their consideration within all EAs. Subsequent guidance and policy statements (FEARO 1994; Hegmann et al. 1999; CEAA 2013) as well as reviews of the methods and practice of cumulative effects assessment (CEA; Duinker et al. 2012) have advanced the science behind and application of the *Canadian Environmental Assessment Act* as well as non-legislative processes designed to address cumulative impacts in ways that are not directly associated with project approval (Harriman and Noble 2008).

C.J. Johnson (✉)
The University of Northern British Columbia, Prince George, BC, Canada
e-mail: Chris.Johnson@unbc.ca



M.P. Gillingham et al. (eds.), *The Integration Imperative*,
DOI 10.1007/978-3-319-22123-6_2

Despite a long history of practice in Canada and even more extensive experience within other jurisdictions (e.g. the USA; see Kenna 2011), there has been considerable criticism of the processes for addressing cumulative impacts (Ross 1998; Noble 2004, 2009; Duinker and Greig 2006; Gunn and Noble 2011). This was well articulated by Duinker et al. (2012, p. 50) in their comprehensive review of the scientific elements and practice of CEA:

> If we do not engage in competent CEA, then the degree to which our activities jeopardize the sustainability of valued ecosystem components will be unknown. Judging that to be an undesirable situation, we conclude that improvements in CEA practice are desperately needed.

The repeated call for improvement in CEA arises from the complexity of defining cumulative impacts. From one perspective, this is an intuitive, easily explained concept that is visibly obvious to most people who live near any of Canada's natural resource industries, or who have an interest in public policy. From a different perspective, that of the industries or governments that are being forced to address cumulative impacts—because of a regulatory response, broad concern about sustainability, land-use conflicts, or control of the use and benefits of natural resources—the problem appears intractable and solutions are elusive. This difficulty arises from our inability to properly quantify impacts or even qualitatively describe the range and scale of the cumulative effects that produce those impacts. The imprecision and uncertainty in identifying cumulative impacts is exacerbated by a decision-making structure, including aspects related to mitigation, that is biased toward socioeconomic benefits (Johnson 2013).

Much of the criticism of current practice is directed at regulatory structures, such as the *Canadian Environmental Assessment Act* (Government of Canada 2012), that are designed to accommodate single proponents and assess individual projects on a site-by-site basis (see Vignette 6.8). Cumulative impacts result from multiple projects that span diverse resource sectors and the impacts can occur regionally over long time periods. Thus, the true cumulative impacts from any one project are difficult for a proponent or even a regulatory agency to consider. Likewise, the calls from some parties to simply limit or identify thresholds for cumulative impacts resulting from many projects are unrealistic given the current focus on economic development and the limitations of EA legislation (AXYS Environmental Consulting 2001; Kennett 2006; Hunter et al. 2009). This is not to say that thresholds are unimportant; an ecosystem that crosses a threshold may begin to degrade rapidly. The problem lies in how to integrate the threshold concept with EA (Johnson 2013).

There has been some work to develop innovative solutions that would be capable of addressing cumulative impacts (Gunn 2009; Gunn et al. 2011). Broad-scale application and testing of these approaches is urgently needed because the cumulative impacts from development are threatening Canada's natural heritage (Fig. 2.1). Nitschke (2008), for example, studied the impacts of 35 years of development across a 410,000-ha area of northeastern BC. He reported changes in the age and structure of forested ecosystems that led to shifts in biodiversity that were not only additive relative to landscape change but also synergistic. Nellemann et al. (2003), Johnson et al. (2005), Boulanger et al. (2012), and Wilson et al. (2012) reported a

0 1 2 4 6 8 km

**Fig. 2.1** A Landsat Thematic Mapper satellite image (1:100,000 scale) illustrates the cumulative effects (i.e. *nibbling loss*) of forest harvesting, road construction, and petroleum exploration and development across an area of central Alberta (Landsat imagery courtesy of the U.S. Geological Survey). These activities will have both positive (e.g. economic development) and negative (e.g. loss and fragmentation of habitat for species that depend on old-growth forests) impacts for the region

large decline in past or predicted future habitat for caribou and reindeer caused by industrial development that is occurring across the central Arctic. These subspecies have high subsistence and cultural value for Canadians, and particularly for Aboriginal communities. Squires et al. (2010), working in the Athabasca River Basin of central Alberta, reported significant cumulative impacts caused by a range of land-use types, including pulp mill effluent, human population growth, agriculture, and oil sands operations. They found significant decreases in water volume and quality (e.g. concentrations of phosphorus, nitrogen, and sulphate) when comparing the most recent data with data from the previous 20 years. Such changes to the natural environment, including the accelerating development of resources, are having real impacts on the health of citizens and the long-term sustainability of communities (Barth 2013; Jeffery et al. 2013; Kinnear et al. 2013; see Vignette 6.7).

The cumulative impacts of human development are now becoming obvious (Fig. 2.1), and policy makers and natural resource professionals are working to develop effective solutions. However, progress in developing better methods for conducting CEA as well as in the planning, policy, and legislation required to address impacts is occurring against a backdrop of increasing industrial activity. In BC, energy development is intensifying in both scope and magnitude. This includes major expansion in the development of coal deposits, petroleum resources, wind energy, and both large- and small-scale hydroelectric facilities (see Box 3.1). An interprovincial committee recently announced a doubling in the known reserves of natural gas, potentially providing a 150-year supply for both domestic use and export (NEB et al. 2013). There are expressions of interest to construct more than five natural gas and two heavy oil pipelines and associated export facilities across coastal BC. The creation of a Ministry of Natural Gas Development in 2013 shows that the province of BC is fully embracing these opportunities as a path to future prosperity. Such twenty-first century developments will occur across landscapes that have a long history of existing and past impacts from forestry, agriculture, mining, and oil and natural gas.

The direct and indirect cumulative impacts of past and present developments have raised concerns among both governments and environmental advocates. More urgent, however, is the requirement to identify the type, rate, and extent of future developments that will not compromise the resilience of ecosystems, nor will compromise the sustainability of communities. As has been witnessed in other jurisdictions, the options for changing the pace of human development decrease over time because of regulatory and tenure inertia and the evolution of a status quo mentality (Timoney and Lee 2001; Aumann et al. 2007). Decisions on land use are a function of the wants of citizens, but are also influenced by market conditions and the transfer of rights to the land in the form of tenures and licences. The window of opportunity for land-use decision making is closing for jurisdictions that are currently developing their natural resources to meet growing global demand. Clearly, there is an urgent need to identify effective methods for understanding and quantifying past and potential future impacts, and then implementing processes for conducting balanced, multi-value decision making.

In this chapter, I explore the fundamental concepts underlying CEA and the range of potential approaches capable of addressing impacts. First, I describe cumulative effects and then differentiate them from cumulative impacts. I continue with a discussion of the rate of development of natural resources in Canada and the relationship between development and the emerging crisis of cumulative impacts. The chapter ends with a review of current methods and approaches for assessing the magnitude and extent of these impacts. This includes current EA legislation and more progressive ideas focused on holistic cumulative effects assessment and management frameworks.

## 2.2 Defining *Cumulative*

The definition of *cumulative* is key to not only understanding and quantifying the scope and magnitude of changes but also developing collaborative frameworks to address the resulting consequences. Although this is an intuitive concept, regulatory agencies, practitioners, and academics have provided multiple definitions of *cumulative impacts* and its consequences (see Duinker et al. 2012 for a review). Common definitions that are quoted from the regulatory processes applied in Canada include:

> ...any cumulative environmental effects that are likely to result from the designated project in combination with other physical activities that have been or will be carried out. (*Canadian Environmental Assessment Act* (Government of Canada 2012), Section 19(1)(a))
>
> Cumulative effects are changes to the environment that are caused by an action in combination with other past, present and future human actions. (Hegmann et al. 1999, p. 3)

These definitions provide a simple, but limited description of how changes accumulate. In particular, they support the common misperception that *cumulative* means *additive*, and that it results from multiple adjacent or overlapping projects. Although the *death by a thousand cuts* metaphor seems appropriate in this context, cumulative processes are much more complex in reality. Effects that originate within one project or across multiple projects may interact or result in nonlinear net consequences that are a product of time-lags or threshold responses. Consistent with the ideas of Ross (1998) and of Harriman and Noble (2008), we suggest a more inclusive definition that considers a fuller range of changes and their consequences:

> *Cumulative* refers to the synergistic, interactive, or unpredictable outcomes of multiple land-use practices or development projects that aggregate over time and space, and that result in significant consequences for people and the environment.

We argue that the definition should consider not only healthy environments but also healthy communities and societies, with the concept of *health* including socioeconomic resilience and self-determination (Parlee et al. 2012). Although human health and socioeconomic well-being depend on naturally functioning and resilient ecosystems, it is necessary to explicitly recognise these human dimensions of the environment.

### 2.2.1 Differentiating Between Effects and Impacts

For the discussions in this book, we have differentiated between the terms *effect* and *impact*: an effect is a change to the environment (including its human components) and an impact represents the consequences of such changes (Wärnbäck and Hilding-Rydevik 2009). Thus, one might quantify the cumulative effects of some set of development activities as the amount of forest that is converted to an early-successional plant and animal community or the increase in the density of linear features in a landscape. The assessment of impacts will depend on how a landscape is perceived as being changed in the short term and the long term by development such as the creation of roads, seismic lines, well pads, mine sites, or clearcuts.

As is the case for effects, impacts are context-specific. During a CEA, one would not aim to identify and quantify all changes to the environment, human health, or communities. Instead, a series of important environmental attributes or values would be identified and the change in those values might be related to the total effect or some subset of the measured effects. Whether a change is positive or negative depends on the values that have been defined, and the impacts (consequences) of changes in the values. Thus, differentiating effects from impacts allows one to identify both positive and negative impacts of cumulative effects. If we consider, for example, the environment, forest harvesting will have an aggregate effect on the amount of old forest. Early-successional forest types might result in a greater number of moose, a species sought after by Aboriginal, recreational, and guided hunters. This would be a positive impact. Likewise, a higher density of moose associated with these forest types might result in a greater number of predators, which would ultimately result in the decline of woodland caribou (see Box 3.2), a conservation species that is currently receiving provincial and federal protection (Serrouya et al. 2011). Similarly, resource development in a town or region might bring employment and higher wages to residents as well as tax revenues for municipal and provincial governments. The negative impacts might include higher housing costs and civil services that can no longer meet the demand of an increasing and potentially transient population (see Chap. 4). For both examples, decision-makers will need to consider the positive and negative impacts of a single development activity or a set of activities.

### 2.2.2 Relating Effects to Impacts

Many EAs are structured according to pathways that link the cause, or effect, and the resulting impact, as defined by some set of valued components (VCs; BC EAO 2011) or valued ecosystem components (VECs; Noble 2010). For the province of BC, a VC is a part of the human or natural environment with ecological, economic, social, cultural, or health importance to a proponent, government, or the public, and that must be recognised and maintained through the EA process. Within the federal assessment process in Canada, a VEC represents a biotic or abiotic component of an ecosystem that has scientific, social, cultural, economic, historical, archaeological, or aesthetic importance. This model requires one to identify the source of an effect and the resulting functional pathway that represents the type of impact or impacts originating from that source. Pathways from multiple sources would reveal an additive or more complex relationship (e.g. interactive, nonlinear) that defines the cumulative impact for the VC or VEC of interest.

Pathways can be useful for defining the mechanistic relationship between sources of effects and the resulting impacts, particularly when describing how these cumulative impacts occur. For the natural resource sector, the most often cited cause of cumulative impacts is what is termed the *nibbling loss*. As the name suggests, this is the additive loss of habitat or some other VC resulting from a cumulative increase in the footprint of human development (Hegmann et al. 1999; Fig. 2.1). *Growth-inducing* effects are more complex to quantify and predict. Here, new development can result in an infrastructure that supports other development that may greatly exceed the cumulative impacts of the first project (Fig. 2.2). Growth-inducing projects include major roads that provide access to new areas or power infrastructure

**Fig. 2.2** The 287 kV Northwest Transmission Line, under construction in this image, will stretch 344 km across an undeveloped but mineral-rich region of northwestern BC (Photo by C. Johnson). The transmission line will provide electricity to planned and future mines as well as an interconnection point for run-of-river hydroelectric development. The right-of-way and associated infrastructure (effects) will have direct impacts on biodiversity and other values. More complex, however, are the indirect economic growth-inducing effects resulting from the provision of inexpensive electricity to large industrial projects. The positive and negative impacts of future resource development will greatly exceed the impacts of the transmission line; however, those impacts are difficult to predict and consider during project assessment and approval

that facilitates energy-dependent industrial activities such as mining. Many projects implemented in a small area over a short time can result in impacts related to a *crowding effect* (Fig. 2.1). The environment may be resilient against some level of activity, but if that level is reached during a too-short period of time, the activity could exceed an ecological or societal threshold for a particular VC. As an example, a body of water may absorb some level of a nutrient or pollutant until a threshold is reached and the water is no longer potable or suitable for a valued population of fish. Similarly, forested landscapes can accommodate some logging, especially when harvest levels and patch sizes are consistent with the natural disturbance regime for that ecosystem (DeLong 2007). However, harvesting beyond that natural regime will fundamentally alter the age-class distribution and size of forest patches in the landscape, creating impacts for plants and animals that depend on these characteristics of the landscape.

## 2.3 The Need to Address Cumulative Impacts: An Emerging Issue

There are considerable challenges in assessing past and current cumulative effects and finding solutions to mitigate their impacts, including the restoration of damaged ecosystems (Duinker et al. 2012). Given projected rates of natural resource development, those challenges cannot be neglected. Not only is the accurate prediction of future impacts difficult, preventing the planning of responses, but there has been very little effort on the part of governments to correct ineffective legislative frameworks. If the expected high rates of development are realised, we are quickly approaching a cumulative impacts crisis.

Canada's economic history was defined by the development and export of natural resources, from beaver pelts to masts for sailing ships. The global economy and government policy ensure that Canada's history continues to determine the country's future, albeit with some diversification. Globally, Canada is ranked fourth in the export of electricity, sixth in crude oil production (third in reserves), fifth in natural gas exports, and ninth in $CO_2$ emissions from energy consumption (CIA 2013). New technologies and practices, combined with expanding export markets, suggest that the development of hydrocarbons will not decrease in the future, but rather will increase. The National Energy Board projects a 75 % increase in Canadian oil production by 2035. In situ oil sands will account for the majority of the new production (*bitumen*; NEB et al. 2013; Fig. 2.3).

Over the same period, the production of natural gas will increase by 25 %—but these are mainly nonconventional sources consisting of tight gas. This gas occurs in rock with low permeability in which extraction requires hydraulic fracturing (fracking) of the rock–gas matrix. Electricity generation is predicted to increase by 27 % (NEB 2013; Fig. 2.4).

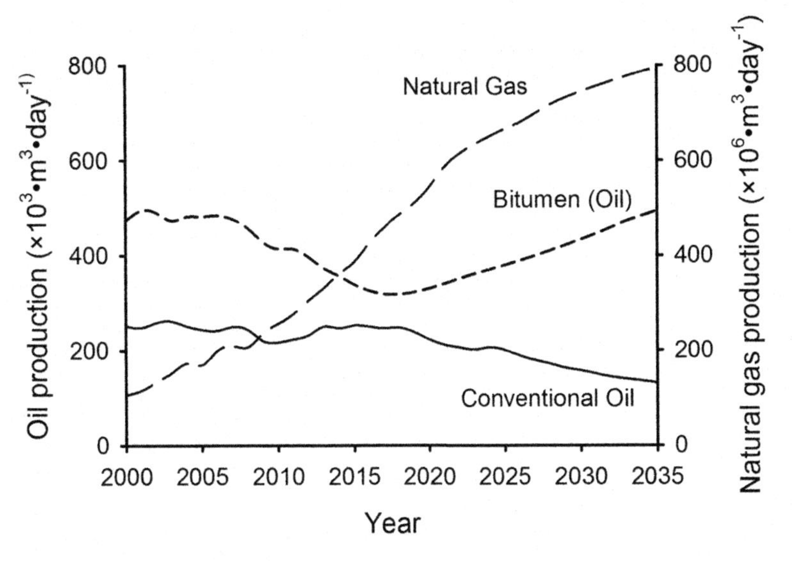


**Fig. 2.3** Projected development of oil and natural gas reserves in Canada (data from NEB et al. 2013)

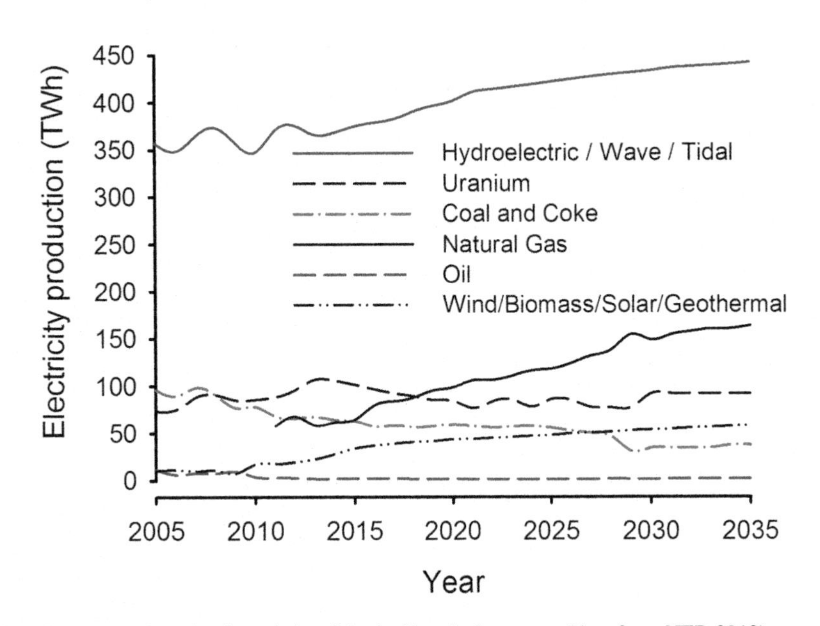


**Fig. 2.4** Projected production of electricity in Canada, by source (data from NEB 2013)

The impacts of individual energy development activities on a given site, such as installation of a pumping station, well, or seismic line, are potentially large, but typically occur over a small area. Of more concern is how the rapid development of conventional and renewable energy resources has the potential to fragment landscapes across extremely large areas, creating impacts for threatened or rare plant and animal species as well as for human communities (Dana et al. 2009; Naugle 2011). McDonald et al. (2009) predicted that the cumulative impacts of future energy development (2009–2030) in the USA would affect more than 20.6 million ha. Across western North America (including BC and Alberta), Copeland et al. (2011) estimated that new and existing energy development could directly or indirectly affect up to 96 million ha. The largest impacts are expected across the boreal forest, followed by shrublands and grasslands.

The cumulative human impacts are also substantial. Air and water pollution reduce environmental quality and thus the quality of the environment for human use (Tenenbaum 2009). Many of these impacts are still not fully understood (Dana et al. 2009). Hydraulic fracturing for natural gas reserves, for example, uses large volumes of water and creates proportionally large amounts of pollution, resulting in unknown long-term consequences for human communities that depend on the affected water resources (Colborn et al. 2011; Souther et al. 2014). From a socio-economic perspective, the rapid growth of emerging centres of energy development can overrun the available municipal and health services (Kinnear et al. 2013). There is also a risk of communities suffering through boom and bust cycles that depend on unpredictable fluctuations in the global demand for local commodities (Barth 2013). In some cases, such dynamics are beyond the control of even the best-meaning and most well-prepared local government (Gramling and Freudenburg 1990).

The alternative to oil, natural gas, and coal, and to their associated impacts, is renewable energy. In 2011, approximately 19 % of global energy consumption was supplied by renewable sources (REN21 2013). The USA and many Canadian provinces have made the further development of renewable energy sources a legislated priority (e.g. *British Columbia Clean Energy Act*; Government of BC 2010). Thus, the development and use of solar, wind, hydropower, biomass, and geothermal energy is likely to grow substantially through the twenty-first century as a response to efforts to reduce $CO_2$ emissions (Fig. 2.4). Even these so-called green sources of energy can, however, result in significant environmental and human impacts (Johnson and Stephens 2011). Wind turbines, for example, are a known cause of mortality for migrating and resident bat and bird populations (Kuvlesky et al. 2007; Pruett et al. 2009). Utility-scale solar energy facilities require large amounts of space and can greatly change the thermal environment of the site (Kaygusuz 2009; Hernandez et al. 2014). For both sources, the required access roads fragment landscapes and further reduce habitat for species that depend on interior forest conditions or undisturbed habitats. The health risks of wind turbines are still being debated, but there are substantive concerns from communities about the impacts of turbine noise and energy transmission on the quality of life of adjacent home owners (Bakker et al. 2012; Jeffery et al. 2013; Groth and Vogt 2014). Development projects in Ontario, for example, have been redesigned or canceled as a result of community concerns and protests. Hydroelectricity has a long history of development in North America, but there are significant impacts for affected river systems, downstream water users, and of course communities that are lost or moved as a response to the impoundment of water (Rosenberg et al. 1997; Zhang and Lou 2011). Micro-hydroelectric projects have a smaller effects footprint, but the distribution of many small facilities leads to a large cumulative impact that results from both the generation sites and the infrastructure for energy and road transportation (Watkin et al. 2011; Bracken and Lucas 2013; Box 3.1).

In addition to energy, Canada's more traditional natural resources (i.e. minerals and forests) are also highly sought after. In 2012, minerals accounted for 21 % of the country's total exports and more than 330,000 Canadians were employed by that resource sector (NRCAN 2012, 2013a; Fig. 2.5). Leading commodities include potash ($7.0 billion), coal ($6.4 billion), gold ($5.6 billion), iron ore ($5.3 billion), copper ($4.5 billion), and diamonds ($2.0 billion). These totals do not include the $3.9 billion spent domestically on exploration and associated costs, with record years in the most recent decade (NRCAN 2012). Globally, Canada is one of the top five producers of aluminium, cadmium, cobalt, nickel, platinum, sulphur, titanium, tungsten, and uranium (USGS 2012). The export of forest products is second only to China, and contributed $25.3 billion to Canada's economy (NRCAN 2013b; Fig. 2.5).

As is the case for energy development, mining and forestry can result in significant cumulative impacts to natural and human systems. This depends, however, on the extent and magnitude of the effects. Forestry can be a sustainable industry if trees are harvested at a rate that allows for regeneration and if a sufficient area of forest is maintained to support biodiversity and other values associated with natural or old-growth forest types (DeLong 2007). Mines may have a relatively small footprint if most of the extraction occurs underground, and the surface can be reclaimed to a more natural state following closure of the mine (Latimer 2012; but see also Raab and Bayley 2012). Both of these industries can, however, have perva-

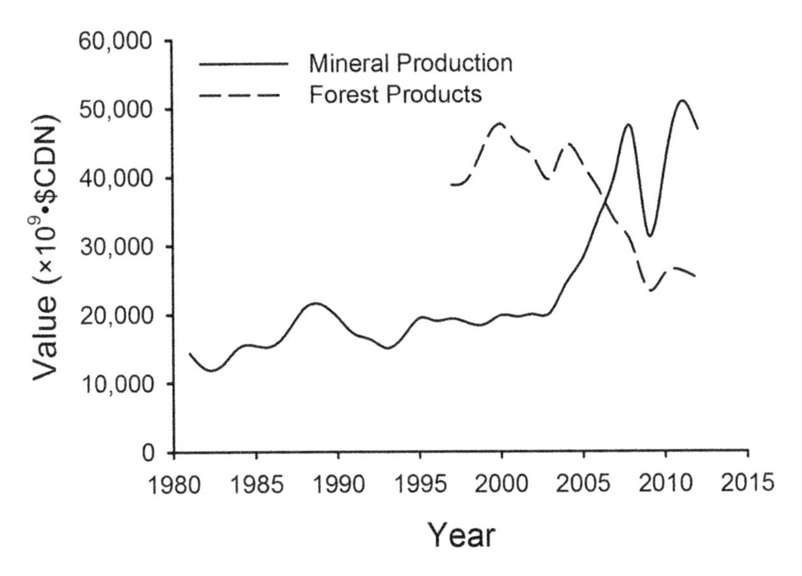


**Fig. 2.5** Economic benefits of mineral production and the export of forest products from Canada (data from NRCAN 2013a, b)

sive cumulative impacts (Fig. 2.1). In particular, forestry results in the development of extensive road networks that may create other impacts, such as the degradation of streams and rivers (Forman and Alexander 1998; Ercelawn 2000; Jones et al. 2000). Mining has a footprint effect with a high magnitude that can exacerbate impacts from adjacent mines or other industrial activities, particularly for surface and near-surface mining. Across the South Peace Region of BC, for example, many areas have had past and recent coal development, as well as oil and natural gas extraction, wind farms, and a long history of forestry. Nitschke (2008) quantified the broad-scale impacts to biodiversity in that region. Woodland caribou, an endangered species across much of BC, have lost considerable amounts of habitat because of the cumulative impacts of those developments (Johnson et al. 2015; see Box 3.2).

Although Canada is a rich country with considerable experience and capacity to regulate the development of natural resources, there is much evidence to recommend improvements in how we address the impacts associated with industrial activities (Timoney and Lee 2001). Nationally, Canada ranks 104th out of 146 countries in its efforts to reduce environmental stresses according to the environmental sustainability index (Esty et al. 2005). When Boyd (2001) compared 25 environmental indicators across the 29 countries in the Organization for Economic Cooperation and Development, Canada ranked 28th. The inherent difficulties in quantifying and managing cumulative impacts and the potential growth in key resource industries suggests that the challenges of maintaining Canada's environments and communities will become more difficult in the future.

The challenges of managing cumulative impacts become even more apparent when considering some of Canada's most vulnerable and disenfranchised communities (Parlee et al. 2012). Across much of the country, Aboriginal peoples are attempting to assert their right to resources and land in the face of past, present, and unprecedented future levels of activity. In BC, these struggles occur in the context of unsettled land claims in which treaty rights or the assertion of Aboriginal title has not yet secured traditional activities or access to twenty-first century resources (see Box 3.6; Vignette 6.6). The BC and federal EA processes do not provide adequate mechanisms for First Nations to have their concerns about cumulative environmental, cultural, and social impacts considered and addressed (FNEMC 2009; Booth and Skelton 2011a). Recognising these deficiencies in legislation and process, Plate et al. (2009) recommended review and improvement of CEA methods, the full inclusion and consideration of oral Aboriginal knowledge when assessing past impacts, and the use of planning or regional assessment processes to set thresholds for these impacts. Booth and Skelton (2011b) also recognised the importance of regional land-use planning or proactive CEAs to support Aboriginal peoples during the review of development projects.

The failure to proactively and meaningfully consider Aboriginal peoples' interests in decision making will undoubtedly result in court challenges that require the provincial and federal governments to reactively address the cumulative impacts of past decisions that now infringe on constitutionally protected Aboriginal rights and title. In BC, there is now precedent for such litigated outcomes. Following a challenge by the West Moberly First Nations, the BC Supreme Court required the province to halt coal development and restore a population of woodland caribou that had declined as a result of cumulative impacts (Box 3.2). This finding may force the province to consult Aboriginal peoples more closely on cumulative impacts when they are reviewing resource development permits and tenures (Findlay and Walton 2010).

## 2.4 Assessing Cumulative Effects and Impacts

Cumulative effects are identified through a broad set of approaches known generically as CEA. The methods are diverse, and VC/VEC specific (see Duinker et al. 2012), but CEA generally refers to a systematic and repeatable approach for assessing the relative strength and significance of cumulative environmental change. Although the term CEA explicitly refers to *effects*, such assessments focus on measuring cumulative environmental impacts (see Sect. 2.2.1). During the assessment process, the results of a CEA are used to determine if a project or series of projects have significant cumulative impacts, and can guide efforts to prevent or mitigate impacts.

Hegmann et al. (1999, p. 3) stated that CEA is "environmental assessment as it should always have been: an Environmental Impact Assessment done well." This speaks to not only the importance of cumulative impacts, relative to the individual impacts from a single project, but also the overwhelming emphasis in Canada on the EA process. As we will discuss in the rest of this chapter, existing regulatory requirements limit the consideration of the full range of impacts as well as the processes that could be developed to address those broader impacts.

A CEA can follow two broad approaches that address either individual projects or the region over which a number of projects have or may occur. The majority of our experience in Canada is with *stressor* or *project-based* CEA (Dubé 2003). These assessments are a response to regulatory review and approval for an individual project and are focused on assessing the cumulative impacts associated with a particular set of effects. These impacts are identified for each VC based on the assumptions that all relevant VCs will be considered and that the mechanisms defining the effect–impact pathways are understood. Also, the range of stressors is confined to the CEA study area for a particular development proposal, thus the spatial and temporal scales of the impacts must be defined carefully. Without such a consideration, it is possible to inadvertently exclude impacts that occur across large regional areas or that become significant over long time periods. For example, chronic health impacts resulting from a project may not be well understood initially, but may become apparent following long-term exposure or through the development of more sensitive diagnostic methods. The primary steps in most project-based CEA (Noble 2010) are:

1. Identify the VCs for a particular project.
2. Determine past, present, and reasonably foreseeable future activities and their associated effects.

3. For each effect, identify the pathway for how it affects the VC and the hypothesised impacts for that VC.
4. Develop a model or method for effect and impact measurement.
5. Quantify the potential impacts of all activities related to the listed VCs.
6. Determine the significance of the impacts.
7. Identify appropriate measures for environmental management and mitigation for all impacts judged to be significant.

In contrast, *effects-based* or *regional* CEA focus on the regional impacts of a series of development projects rather than single stressors or project-specific effects and their associated impacts. This shift in both method and philosophy arises from the recognition that CEA should extend beyond the scope of any one project or proponent (Dubé 2003). Regional CEA provides greater flexibility to consider broader spatial and temporal scales and a wider scope of VCs, and is less concerned with the potentially complex mechanisms embodied in the effect–impact pathways (Noble 2010). Regional studies are more closely aligned with broad-scale sustainability targets. Thus, regional CEA is flexible and responsive to regional needs, and facilitates approaches for cumulative impacts management that engage broader communities and that emphasise proactive planning (Johnson 2011).

Regional studies of cumulative impacts are recognised and even *encouraged* within the *Canadian Environmental Assessment Act* (S.4(1); Government of Canada 2012). The dominant model for CEA, however, remains project-based. This is likely the result of a lack of government leadership, since no one industry or proponent can be expected to plan for regional impacts unrelated to their activities. Also, regulations are a key driver of EA, but necessarily focus on projects and proponents.

## 2.5 Regulatory Requirements for Cumulative Effects Assessment

The cumulative impacts of resource development are one of the largest emerging challenges for natural resource, health, and planning professionals as well as for communities that are attempting to accommodate rapidly expanding industrial sectors (Krausman and Harris 2011; Naugle 2011; Kinnear et al. 2013). In addition, there is an expanding scope of values and increasing appreciation of the holistic nature and complexity of ecological systems (see Chap. 3). Ecosystem services, for example, are now recognised as more than an abstract set of metrics that can be used to measure environmental change; they are increasingly seen as real services that influence the health and well-being of humans (Costanza et al. 1997; Carpenter et al. 2009). In reality, however, much of the contemporary work that has focused on understanding cumulative impacts, including the application and critiques of CEA, has occurred in response to EA legislation (Dixon and Montz 1995; Duinker and Grieg 2006). Much of that legislation has focused exclusively on a subset of discrete environmental values, with socioeconomic and health impacts a consideration only when they relate directly to environmental change.

Legislative requirements for CEA vary across Canada. The *Canadian Environmental Assessment Act* (Government of Canada 2012) is federal legislation that applies to all projects that meet specific requirements relative to the involvement and responsibilities of the federal government. This includes projects: that affect fish and fish habitat, migratory birds and their habitat; that have cross-border effects (provincial and national) or that affect Aboriginal people; that occur on federal land or cause changes to the environment as a result of a decision by the federal government; and projects that are regulated by the Canadian Nuclear Safety Commission or the National Energy Board.

Cumulative effects assessment may vary considerably among projects reviewed as a requirement of the *Canadian Environmental Assessment Act* (Government of Canada 2012). Criteria for defining the approach and level of effort include (CEAA 2013):

- The type of project;
- The magnitude of the anticipated potential cumulative impacts;
- The health or status of VECs that may be affected by the cumulative environmental effects;
- The potential for mitigation; and
- The level of concern expressed by Aboriginal groups or the general public.

Most provinces and territories in Canada have separate legislation that considers the environmental and socioeconomic impacts of resource development proposals. Some of these legislative frameworks have provisions for assessing the cumulative impacts. For example, the *BC Environmental Assessment Act* (Government of BC 2002) was amended in 2010 to formally recognise cumulative effects. Proponents can voluntarily consider these cumulative effects, but whether this is required is at the government's discretion. There have therefore been examples of projects that included a CEA, but there is no formal guidance on the scope or type of CEA that is acceptable to the BC Environmental Assessment Office (Haddock 2010).

In Ontario, cumulative effects are considered as a matter of policy rather than explicitly within the context of legislation. According to the Ontario Ministry of the Environment's *Statement of Environmental Values*, the Ministry must: "…consider the cumulative effects on the environment; the interdependence of air, land, water and living organisms; and the relationships among the environment, the economy and society" (Government of Ontario 2014). In Alberta, the *Land Stewardship Act* (Government of Alberta 2009) provides a statutory framework by which the provincial government can develop regional plans that manage cumulative effects. Planning is occurring across seven broad regions and includes a consideration of the impacts associated with water quality and supply, pollutant emissions, and habitat loss. Plans are approved by the provincial Cabinet and are meant to provide policy direction for the regions, thereby supporting finer-scale decision making.

Although there are legislative tools at various jurisdictional levels to consider cumulative impacts, and although these sometimes require formal CEA, much of the emphasis is placed on measuring and addressing changes to the environment resulting from individual projects. The *Canadian Environmental Assessment Act* (Government of Canada 2012), for example, focuses exclusively on the environment. There is no consideration of socioeconomic or human health impacts that are not a direct function of some change in the environment caused by project activities. In contrast, the BC Environmental Assessment Office often considers a fuller range of impacts, including the economic benefits of a project (BC EAO 2011; Pockey 2011). One must be cautious, however, when tax benefits and employment statistics become the standard by which impacts are measured. Where environmental change, community resilience, and human health cannot be assigned monetary value, the consideration of project revenues may result in underestimation of the negative impacts to less tangible or quantifiable VCs. As Hegmann et al. (1999) recognised, there needs to be a better consideration of and more effective methods for considering the influence of environmental impacts on socioeconomic systems as well as the impacts of cumulative socioeconomic changes on the environment (see Chap. 4). We suggest that such improvements in the process are also required to better consider the impacts of development, including environmental change, on human health (see Chap. 5).

## 2.6 Alternatives to Regulation

There have been strong criticisms of the effectiveness and even the role of project-specific EA in efforts to address cumulative impacts (Burris and Canter 1997; Baxter et al. 2001). Many have argued that cumulative impacts are inadequately represented in existing legislative frameworks or, at a more fundamental level, not served well by the structure and application of the EA process (Creasey 1998; Kennett 1999; Davey et al. 2002). Cumulative impacts are not immediately associated with the time and place of a proposed development and, therefore, it may be difficult to define the extent or magnitude of an impact (McCold and Saulsbury 1996). With the exception of regional studies, which are acceptable under the *Canadian Environmental Assessment Act* (Government of Canada 2012), there is no requirement for a strategic vision that would encompass spatial and temporal domains that exceed the footprint of the proposal that triggered the assessment. Furthermore, EA in Canada and beyond (Dixon and Montz 1995), including the formal requirement of a CEA, is a reactive proponent-driven process. An EA considers the impacts of individual projects rather than multiple projects that may span large areas, jurisdictional boundaries, and considerable time periods. The CEA is a secondary consideration and occurs only after direct project impacts are considered.

These are not only issues for regulators and concerned citizens. Often, a proponent will view the process of developing a meaningful CEA as intractable. Even within certain industrial sectors, there is little sharing of the strategic business interests, data, and knowledge that would make such an assessment possible. This problem is magnified by cross-sectoral gaps in communication and relationships. Furthermore, working with various levels of government, including Aboriginal peoples, can be challenging if a transparent and consistent decision-making process is the goal. Each of these governments can bring unique interpretations of the significance of impacts or even the importance of VCs and the bounds of the study area (Pockey 2011). Overlapping assessments and permit-granting processes further complicate project review, and this has obvious costs for businesses and uncertainty for resource-dependent communities.

Many have recommended a tiered decision-making framework as a solution for the current failings of EA in Canada (Conacher 1994; Creasey 1998; Kennett 1999; Davey et al. 2002; Duinker and Greig 2006; Gunn 2009). Such a framework would be implemented at a regional scale and would evaluate current levels of cumulative impacts using standardised metrics that are consistent across resource sectors. Linked to this understanding of regional impacts would be targets for acceptable or desirable levels of future development. For Aboriginal peoples, the development of a strategic vision of development for their traditional territories would lead to more inclusive and effective involvement in the assessment process (Plate et al. 2009). Targets could be based on any number of environmental, socioeconomic, cultural, or health criteria. Thus, the environmental impacts of individual projects could be considered within the context of existing and future impacts. Such a framework would have many of the qualities of effective strategic land-use planning (Booth and Halseth 2011). As noted by Bardecki (1990, p. 322), "Assessing and managing cumulative impacts is planning."

### *2.6.1 Regional Environmental Assessment*

Regional environmental assessment (REA), also referred to as regional CEA, and in Canada, as Regional Strategic Environmental Assessment (RSEA), is a general type of accounting and guidance framework that would accommodate and inform individual project approvals within a broader and more holistic understanding of current and acceptable levels of impacts (Conacher 1994; Bonnell and Storey 2000; Gunn 2009). Regional environmental assessment is recognised globally as well as within the *Canadian Environmental Assessment Act* (Government of Canada 2012). The World Bank (1999) defines REA as:

> An instrument that examines environmental issues and impacts associated with a particular strategy, policy, plan, or program, or with a series of projects for a particular region (e.g., an urban area, a watershed, or a coastal zone); evaluates and compares the impacts against those of alternative options; assesses legal and institutional aspects relevant to the issues and impacts; and recommends broad measures to strengthen environmental management in the region. Regional EA pays particular attention to potential cumulative impacts of multiple activities.

Compared to individual project assessments, REA is outward-looking and strategic, and considers a range of interacting impacts across a region (Harriman and Noble 2008). Failure to look beyond a single development project limits our ability to address the deficiencies of current project-focused EA, and our ability to develop decision-making frameworks that consider cumulative impacts in all of their forms (Baxter et al. 2001; Duinker and Greig 2006).

The benefits of REA are numerous and include: long-term development targets or plans within the context of sustainability; participation of all regulatory agencies and stakeholder groups; identification of a range of environmental effects and impacts early in the land-use decision-making process; assessment of baseline conditions and data gaps; and development of monitoring and management frameworks that support the documentation of explicitly regional cumulative impacts and the significance of the impacts associated with specific projects (Kennett 1999; Davey et al. 2002; Gunn and Noble 2009a). Regional environmental assessment has been proposed or has shown some success for a number of resource development sectors, and there are a range of technical approaches for understanding large-scale cumulative impacts, especially from the perspective of animal and plant communities (Schneider et al. 2003). Some of the failures of REA result from the unrealised expectation of a one-size-fits-all model. Regional environmental assessment is most successful when it is developed to suit the challenges of land use and development in a specific region (Harriman and Noble 2008).

### 2.6.2 *Cumulative Effects Assessment and Management Frameworks*

Some Canadian jurisdictions, recognising the value of integrated project-specific EA and the principles of REA, have developed what are generically known as cumulative effects assessment and management frameworks (CEAMFs). Such frameworks are defined as "an administrative structure that can help decision-makers assess and manage the effects of human use of the land" (AXYS Environmental Consulting 2003, p. 1–6). Cumulative effects assessment and management frameworks are flexible and adaptable to a region's specific challenges related to cumulative impacts. One could therefore consider CEAMFs as the operational realisation of REAs. Thus, past and current experiences with CEAMFs not only provide lessons for better addressing cumulative impacts but perhaps provide templates for future efforts in other regions of Canada.

Gunn and Noble (2009b) identified and reviewed four Canadian CEAMFs. They concluded that the origin, goals, and development of each framework were unique, but there were some common themes, including land-use planning, development of a vision, coordination among regulatory agencies, policy development, and monitoring of cumulative impacts. The Northwest Territories (NWT) CEAMF, one of the first Canadian frameworks, is a good example of both the potential and failings of this approach. Formed in the late 1990s in response to the rapid development of the diamond mining industry, the NWT CEAMF was composed of a steering committee with representation from the territorial, federal, and First Nations governments and councils, and from non-governmental and industry organisations. The Committee was tasked with making recommendations or providing *refusable advice* to decision-makers on a broad list of initiatives that encompassed ecological integrity, sustainable communities, and economic development. Although the NWT CEAMF Implementation Blueprint identified baseline studies, research, and monitoring as necessary components of cumulative effects management, there was little progress in this direction (NWT CEAM Steering Committee 2007). Slow progress on such goals reduced the overall legitimacy of the framework (Gunn and Noble 2009b).

Although the attributes of each framework were unique, Gunn and Noble's (2009b) research identified some common themes that can predict the success or failure of this approach. First, a stakeholder-defined regional vision for future development was important for success. This was consistent with broad spatial and temporal perspectives on cumulative impacts that engaged a range of land-use sectors and their associated stakeholders. Second, members of the frameworks often had difficulties linking the strategic nature of CEAMFs to regulatory decisions, and this difficulty was a predictor of failure. Third, translating strategic visions into operational guidance and tracking progress toward meeting goals was difficult, especially when participants were positioned in agencies tasked with project-level decisions.

Despite these difficulties, CEAMFs provide a real opportunity for conducting CEA at meaningful scales and for structuring processes that would guide land-use planning and site-specific decision making for situations that go beyond individual project proposals. These frameworks will have particular value when directed at hotspots where cumulative impacts are especially severe or are expected to occur in the future. Ultimately, however, the limits of past frameworks will need to be addressed. A more complete and fully realised integration of project-based and regional CEA is the starting point. As part of the path that leads to this integration, governments will need to elevate CEAMFs beyond advisory roles and provide them with some legislated authority to influence land-use decision making.

## 2.7 Conclusions

There are profound changes on the horizon for regions of Canada that are hoping to maintain functioning and resilient ecosystems as well as a high quality of life in the context of an accelerating twenty-first century economy (Parlee et al. 2012). These challenges are especially acute for Aboriginal communities who have been disenfranchised by top-down government processes and who may have culturally unique concerns and solutions for managing industrial development. Similar conclusions can be drawn for other regions of the world where rapid resource development is occurring or expected (McDonald et al. 2009; Copeland et al. 2011). Unfortunately, past approaches to address cumulative impacts appear to have been woefully inadequate, and this is not just the Canadian experience (Dixon and Montz 1995; Burris and Canter 1997). Twenty years of retrospective analysis has demonstrated that project-specific approaches for assessing and addressing cumulative impacts are not sufficient, as they cannot meet the basic principles of sustainability: healthy environments, productive economies, and communities that support a high quality of life (Duinker et al. 2012). The reasons for failure are many, but ultimately result from an assessment process that struggles to look beyond the impacts associated with only a single proposed development in isolation from other development projects (see Chap. 7). The principles of REA offer a starting point for fully considering cumulative impacts across regions and longer (strategic rather than tactical) time periods (see Chap. 8). In Canada, these ideas have been exemplified by CEAMFs. Although these frameworks have not always been successful, and have most often been limited to advisory roles with no authority, at the very least these frameworks provide a mechanism to consider regional scales, multiple resource sectors, and broader participation in land-decision making.

## References


Aumann, C., D.R. Farr, and S. Boutin. 2007. Multiple use, overlapping tenures, and the challenge of sustainable forestry in Alberta. *Forestry Chronicle* 83: 642–650.

AXYS Environmental Consulting. 2001. *Thresholds for addressing cumulative effects on terrestrial and avian wildlife in the Yukon*. Whitehorse: Department of Indian and Northern Affairs and Environment Canada.

———. 2003. *A cumulative effects assessment and management framework (ceamf) for northeastern British Columbia: Volume 1*. Fort St. John: BC Oil and Gas Commission and the Muskwa–Kechika Advisory Board.

Bakker, R.H., E. Pedersen, G.P. van den Berg, R.E. Stewart, W. Lok, and J. Bouma. 2012. Impact of wind turbine sound on annoyance, self-reported sleep disturbance and psychological distress. *Science of the Total Environment* 425: 42–51.

Bardecki, M.J. 1990. Coping with cumulative impacts: An assessment of legislative and administrative mechanisms. *Impact Assessment Bulletin* 8: 319–344.

Barth, J.M. 2013. The economic impact of shale gas development on state and local economies: Benefits, costs, and uncertainties. *New Solutions* 23: 85–101.

Baxter, W., W.A. Ross, and H. Spaling. 2001. Improving the practice of cumulative effects assessment in Canada. *Impact Assessment and Project Appraisal* 19: 253–262.

BC EAO (British Columbia Environmental Assessment Office). 2011. *Environmental assessment office user guide*. Victoria: British Columbia Environmental Assessment Office.

Bonnell, S., and K. Storey. 2000. Addressing cumulative effects through strategic environmental assessment: A case study of small hydro development in Newfoundland, Canada. *Journal of Environmental Assessment Policy and Management* 2: 477–499.

Booth, A., and G. Halseth. 2011. Why the public thinks natural resources public participation processes fail: A case study of British Columbia communities. *Land Use Policy* 28: 898–906.

Booth, A.L., and N.W. Skelton. 2011a. Industry and government perspectives on First Nations' participation in the British Columbia environmental process. *Environmental Assessment Impact Review* 31: 216–225.

———. 2011b. Improving First Nations' participation in environmental assessment processes: Recommendations from the field. *Impact Assessment and Project Appraisal* 29: 49–58.

Boulanger, J., K.G. Poole, A. Gunn, and J. Wierzchowski. 2012. Estimating the zone of influence of industrial developments on wildlife: A migratory caribou and diamond mine case study. *Wildlife Biology* 18: 164–179.

Boyd, D.R. 2001. *Canada vs the OECD: An environmental comparison*. Victoria: Eco-Research Chair of Environmental Law and Policy, University of Victoria.

Bracken, F.S.A., and M.C. Lucas. 2013. Potential impacts of small-scale hydroelectric power generation on downstream moving lampreys. *River Research and Applications* 29: 1073–1081.

Burris, R.K., and L.W. Canter. 1997. Cumulative impacts are not properly addressed in environmental assessments. *Environmental Impact Assessment Review* 17: 5–18.

Carpenter, S.R., H.A. Mooney, J. Agard, D. Capistrano, R.S. DeFries, S. Diaz, T. Dietz, A.K. Duraiappahh, A. Oteng-Yeboahi, H. Miguel Pereiraj, C. Perringsk, W.V. Reidl, J. Sarukhanm, R.J. Scholesn, and

A. Whyte. 2009. Science for managing ecosystem services: Beyond the Millennium Ecosystem Assessment. *Proceedings of the National Academy of Sciences of the United States of America* 106: 1305–1312.

CEAA (Canadian Environmental Assessment Agency). 2013. *Operational Policy Statement—Addressing Cumulative Environmental Effects under the Canadian Environmental Assessment Act, 2012*. Ottawa: Canadian Environmental Assessment Agency.

CIA (Central Intelligence Agency). 2013. The world factbook: Canada. https://www.cia.gov/library/publications/the-world-factbook/. Accessed 1 Dec 2013.

Colborn, T., C. Kwiatkowski, K. Schultz, and M. Bachran. 2011. Natural gas operations from a public health perspective. *Human and Ecological Risk Assessment* 17: 1039–1056.

Conacher, A.J. 1994. The integration of land-use planning and management with environmental impact assessment: Some Australian and Canadian perspectives. *Impact Assessment* 4: 347–373.

Copeland, H.E., A. Pocewicz, and J.M. Kiesecker. 2011. Geography of energy in western North America: Potential impacts to terrestrial ecosystems. In *Energy development and wildlife conservation in North America*, ed. D.E. Naugle, 7–22. Washington: Island Press.

Costanza, R., R. d'Arge, R. de Groot, S. Farber, M. Grasso, B. Hannon, K. Limburg, S. Naeem, R.V. O'Neill, J. Paruelo, R.G. Raskin, P. Sutton, and M. van den Belt. 1997. The value of the world's ecosystem services and natural capital. *Nature* 387: 253–260.

Creasey, J.R. 1998. Cumulative effects and the wellsite approval process. MSc Thesis, University of Calgary.

Dana, L.P., R.B. Anderson, and A. Meis-Mason. 2009. A study of the impact of oil and gas development on the Dene First Nations of Sahtu (Great Bear Lake) region of the Northwest Territories (NWT). *Journal of Enterprising Communities: People and Places in the Global Economy* 3: 94–117.

Davey, L.H., J.L. Barnes, C.L. Horvath, and A. Griffiths. 2002. Addressing cumulative environmental effects: Sectoral and regional environmental assessment. In *Cumulative environmental effects management: Tools and approaches*, ed. A.J. Kennedy, 187–205. Calgary: Alberta Society of Professional Biologists.

DeLong, S.C. 2007. Implementation of natural disturbance-based management in northern British Columbia. *Forestry Chronicle* 83: 338–346.

Dixon, J., and B.E. Montz. 1995. From concept to practice: Implementing cumulative impact assessment in New Zealand. *Environmental Management* 19: 445–456.

Dubé, M.G. 2003. Cumulative effect assessment in Canada: A regional framework for aquatic ecosystems. *Environmental Impact Assessment Review* 23: 723–745.

Duinker, P.N., and L.A. Greig. 2006. The impotence of cumulative effects assessment in Canada: Ailments and ideas for redeployment. *Environmental Management* 37: 153–161.

Duinker, P.N., E.L. Burbidge, S.R. Boardley, and L.A. Greig. 2012. Scientific dimensions of cumulative effects assessment: Toward improvements in guidance for practice. *Environmental Review* 21: 40–52.

Ercelawn, A. 2000. *End of the road: the adverse ecological impacts of roads and logging: A compilation of independently reviewed research*. New York: Natural Resources Defence Council.

Esty, D.C., M. Levy, T. Srebotnjak, and A. de Sherbinin. 2005. *Environmental sustainability index: Benchmarking national environmental stewardship*. New Haven: Yale Center for Environmental Law and Policy.

FEARO (Federal Environmental Assessment Review Office). 1994. *A reference guide for the Canadian environmental assessment act: Addressing cumulative environmental effects*. Hull: Federal Environmental Assessment Review Office.

Findlay, C., and J. Walton. 2010. Coal, caribou and consultation: Court directs accommodation in Treaty 8 region. Blakes Bulletins: Blake, Cassels and Graydon LLP. http://www.blakes.com/English/Resources/Bulletins/Pages/Details.aspx?BulletinID=1115. Accessed 26 March 2015.

FNEMC (First Nations Energy and Mining Council). 2009. *Environmental assessment and First Nations in BC: Proposals for reform*. North Vancouver: First Nations Energy and Mining Council.

Forman, R.T.T., and L.E. Alexander. 1998. Roads and their major ecological effects. *Annual Review of Ecology and Systematics* 29: 207–231.

Government of Alberta. 2009. Land Stewardship Act, 2009 (S.A. 2009, C. A-26.8). http://www.qp.alberta.ca/documents/Acts/A26P8.pdf. Accessed 10 Dec 2014.

Government of BC. 2002. Environmental Assessment Act, 2002 (S.B.C. 2002, C. 43). http://www.bclaws.ca/civix/document/id/complete/statreg/02043._01 Accessed 10 Dec 2014.

———. 2010. Clean Energy Act, 2010 (S.B.C. 2010, C. 22). http://www.bclaws.ca/civix/document/id/complete/statreg/10022_01. Accessed 12 Dec 2014.

Government of Canada. 2012. Canadian Environmental Assessment Act, 2012 (S.C. 2012, C. 19, S. 52). http://laws-lois.justice.gc.ca/eng/acts/c-15.21/page-1.html. Accessed 12 Dec 2014.

Government of Ontario. 2014. Statement of Environmental Values: Ministry of Environment. www.ebr.gov.on.ca/ERS-WEB-External/content/sev.jsp?pageName=sevList&subPageName=10001. Accessed 15 Dec 2014.

Gramling, R., and W.R. Freudenburg. 1990. A closer look at "local control": Communities, commodities and the collapse of the coast. *Rural Sociology* 55: 541–558.

Groth, T.M., and C.A. Vogt. 2014. Rural wind farm development: Social, environmental and economic features important to local residents. *Renewable Energy* 63: 1–8.

Gunn, J.H. 2009. Integrating strategic environmental assessment and cumulative effects assessment in Canada. PhD Dissertation, University of Saskatchewan.

Gunn, J.H., and B.F. Noble. 2009a. A conceptual basis and methodological framework for regional strategic environmental assessment (R-SEA). *Impact Assessment and Project Appraisal* 27: 258–270.

———. 2009b. Integrating cumulative effects in regional strategic environmental assessment frameworks: Lessons from practice. *Journal of Environmental Assessment Policy and Management* 11: 267–290.

———. 2011. Conceptual and methodological challenges to integrating SEA and cumulative effects assessment. *Environmental Impact Assessment Review* 31: 154–160.

Gunn, A., C.J. Johnson, J.S. Nishi, C.J. Daniel, D.E. Russell, M. Carlson, and J.Z. Adamczewski. 2011. Understanding the cumulative effects of human activities on barren-ground caribou. In *Cumulative effects in wildlife management: Impact mitigation*, ed. P.R. Krausman and L.K. Harris, 113–133. Boca Raton: CRC Press.

Haddock, M. 2010. *Environmental assessment in British Columbia*. Victoria: Environmental Law Centre, University of Victoria.

Harriman, J.A., and B.F. Noble. 2008. Characterizing project and strategic approaches to regional cumulative effects assessment in Canada. *Journal of Environmental Assessment Policy and Management* 10: 25–50.

Hegmann, G., C. Cocklin, R. Creasey, S. Dupuis, A. Kennedy, L. Kingsley, W. Ross, H. Spaling, and D. Stalker. 1999. *Cumulative effects assessment practitioners guide*. Hull: AXYS Environmental Consulting, and the CEA Working Group for the Canadian Environmental Assessment Agency.

Hernandez, R.R., S.B. Easter, M.L. Murphy-Mariscal, F.T. Maestre, M. Tavassoli, E.B. Allen, C.W. Barrows, J. Belnap, R. Ochoa-Hueso, S. Ravi, and M.F. Allen. 2014. Environmental impacts of utility-scale solar energy. *Renewable and Sustainable Energy Reviews* 29: 766–779.

Hunter, M.L., M.J. Bean, D.B. Lindenmayer, and D.S. Wilcove. 2009. Thresholds and the mismatch between environmental laws and ecosystems. *Conservation Biology* 23: 1053–1055.

Jeffery, R.D., C. Krogh, and B. Horner. 2013. Adverse health effects of industrial wind turbines. *Canadian Family Physician* 59: 473–475.

Johnson, C.J. 2011. Regulating and planning for cumulative effects: The Canadian experience. In *Cumulative effects in wildlife management: Impact mitigation*, ed. P.R. Krausman and L.K. Harris, 29–46. Boca Raton: CRC Press.

———. 2013. Identifying ecological thresholds for regulating human activity: Effective conservation or wishful thinking? *Biological Conservation* 168: 57–65.

Johnson, G.D., and S.E. Stephens. 2011. Wind power and biofuels: A green dilemma for wildlife conservation. In *Energy development and wildlife conservation in North America*, ed. D.E. Naugle, 131–157. Washington: Island Press.

Johnson, C.J., M.S. Boyce, R.L. Case, H.D. Cluff, R.J. Gau, A. Gunn, and R. Mulders. 2005. Quantifying the cumulative effects of human developments: A regional environmental assessment for sensitive Arctic wildlife. *Wildlife Monographs* 160: 1–36.

Johnson, C.J., L.P.W. Ehlers, and D.R. Seip. 2015. Witnessing extinction: Cumulative impacts across landscapes and the future loss of an evolutionarily significant unit of woodland caribou in Canada. *Biological Conservation* 186: 176–186.

Jones, J.A., F.J. Swanson, B.C. Wemple, and K.A. Snyder. 2000. Effects of roads on hydrology, geomorphology, and disturbance patches in stream networks. *Conservation Biology* 14: 76–85.

Kaygusuz, K. 2009. Environmental impacts of the solar energy system. *Energy Sources, Part A: Recovery, Utilization, and Environmental Effects* 31: 1376–1386.

Kenna, M. 2011. The NEPA process: What the law says. In *Cumulative effects in wildlife management: Impact mitigation*, ed. P.R. Krausman and L.K. Harris, 17–27. Boca Raton: CRC Press.

Kennett, S.A. 1999. *Towards a new paradigm for cumulative effects management*. Calgary: Canadian Institute of Resources Law, University of Calgary. Occasional Paper #8.

Kennett, S. 2006. *From science-based thresholds to regulatory limits: Implementation issues for cumulative effects management*. Yellowknife: Environment Canada.

Kinnear, S., Z. Kabir, J. Mann, and L. Bricknell. 2013. The need to measure and manage the cumulative impacts of resource development on public health: An Australian perspective. In *Current topics in public health*, ed. A. Rodriguez-Morales, 125–144. Rijeka: InTech Publishers.

Krausman, P.R., and L.K. Harris (eds.). 2011. *Cumulative effects in wildlife management: Impact mitigation*. Boca Raton: CRC Press.

Kuvlesky, W.P., L.A. Brennan, M.L. Morrison, K.K. Boydston, B.M. Ballard, and F.C. Bryant. 2007. Wind energy development and wildlife conservation: Challenges and opportunities. *Journal of Wildlife Management* 71: 2487–2498.

Latimer, C.E. 2012. Avian population and community dynamics in response to vegetation restoration on reclaimed mine lands in southwest Virginia. MSc Thesis, Virginia Polytechnic Institute and State University.

McCold, L.N., and J.W. Saulsbury. 1996. Including past and present impacts in cumulative impact assessments. *Environmental Management* 20: 767–776.

McDonald, R.I., J. Fargione, J. Kiesecker, W.N. Miller, and J. Powell. 2009. Energy sprawl or energy efficiency: Climate policy impacts on natural habitat for the United States of America. *PLoS One* 4: e6802.

Naugle, D.E. (ed.). 2011. *Energy development and wildlife conservation in North America*. Washington: Island Press.

NEB (National Energy Board). 2013. *Canada's energy future: Energy supply and demand projections to 2035*. Ottawa: National Energy Board. https://www.neb-one.gc.ca/nrg/ntgrtd/ftr/index-eng.html. Accessed 1 Dec 2013.

NEB (National Energy Board), BC Oil and Gas Commission, Alberta Energy Regulator, and BC Ministry of Natural Gas Development. 2013. *Energy briefing note: The ultimate potential for unconventional petroleum from the Montney formation of British Columbia and Alberta*. Calgary: National Energy Board, BC Oil and Gas Commission, Alberta Energy Regulator, and BC Ministry of Natural Gas Development.

Nellemann, C., I. Vistnes, S.O. Jordhøy, and A. Newton. 2003. Progressive impact of piecemeal infrastructure development on wild reindeer. *Biological Conservation* 113: 307–317.

Nitschke, C.R. 2008. The cumulative effects of resource development on biodiversity and ecological integrity in the Peace-Moberly region of Northeast British Columbia, Canada. *Biodiversity and Conservation* 17: 1715–1740.

Noble, B.F. 2004. A state-of-practice survey of policy, plan, and program assessment in Canadian provinces. *Environmental Impact Assessment Review* 24: 351–361.

———. 2009. Promise and dismay: The state of strategic environmental assessment systems and practices in Canada. *Environmental Impact Assessment Review* 29: 66–75.

———. 2010. *Introduction to environmental impact assessment: A guide to principles and practice*, 2nd ed. Don Mills: Oxford University Press.

NRCAN (Natural Resources Canada). 2012. Key Mining Facts 2012. http://www.nrcan.gc.ca/minerals-metals/nmw-smc/4450. Accessed 1 Dec 2013.

———. 2013a. Mineral Trade Information Bulletin 2013. http://www.nrcan.gc.ca/minerals-metals/publications-reports/3264. Accessed 1 Dec 2013.

———. 2013b. Selective cuttings. http://cfs.nrcan.gc.ca/selective-cuttings/54. Accessed 1 Dec 2013.

NWT CEAM (Northwest Territories Cumulative Effects Assessment and Management) Steering Committee. 2007. *A blueprint for implementing the cumulative effects assessment and management strategy and framework in the NWT and its regions*. Yellowknife: NWT CEAM Steering Committee.

Parlee, B.L., K. Geertsma, and A. Willier. 2012. Social-ecological thresholds in a changing boreal landscape: Insights from Cree knowledge of the Lesser Slave Lake Region of Alberta, Canada. *Ecology and Society* 17(2): 20.

Peterson, E.B., Y.H. Chan, N.M. Peterson, G.A. Constable, R.B. Caton, C.S. Davis, R.R. Wallace, and G.A. Yarronton. 1987. *Cumulative effects assessment in Canada: An agenda for action and research*. Hull: Canadian Environmental Assessment Research Council.

Plate, E., M. Foy, and R. Krehbiel. 2009. *Best practices for First Nation involvement in environmental assessment reviews of development projects in British Columbia*. West Vancouver: New Relationship Trust.

Pockey, M. 2011. Lessons learned from the prosperity mine decision: Enhancing project certainty through a social licence strategy. *Environment and Energy Bulletin* 3: 1–8.

Pruett, C.L., M.A. Patten, and D.H. Wolfe. 2009. It's not easy being green: Wind energy and a declining grassland bird. *Bioscience* 59: 257–262.

Raab, D., and S.E. Bayley. 2012. A vegetation-based index of biotic integrity to assess marsh reclamation success in Alberta oil sands, Canada. *Ecological Indicators* 15: 43–51.

REN21. 2013. Renewables 2013 Global Status Report. Paris: REN21 Secretariat. http://www.ren21.net/portals/0/documents/resources/gsr/2013/gsr2013_lowres.pdf. Accessed 6 Feb 2015.

Retief, F., C. Jones, and S. Jay. 2009. The emperor's new clothes—reflections on strategic environmental assessment (SEA) practice in South Africa. *Environmental Impact Assessment Review* 28: 504–514.

Rosenberg, D.M., F. Berkes, R.A. Bodaly, R.E. Hecky, C.A. Kelly, and J.W.M. Rudd. 1997. Large-scale impacts of hydroelectric development. *Environmental Reviews* 5: 27–54.

Ross, W.A. 1998. Cumulative effects assessment: Learning from Canadian case studies. *Impact Assessment and Project Appraisal* 16: 267–276.

Samarakoon, M., and J.S. Rowan. 2008. A critical review of environmental impact statements in Sri Lanka with particular reference to ecological impact assessment. *Environmental Management* 41: 441–460.

Schneider, R.R., J.B. Stelfox, S. Boutin, and S. Wasel. 2003. Managing the cumulative impacts of land uses in the Western Canadian Sedimentary Basin: A modelling approach. *Conservation Ecology* 7(1): 8.

Serrouya, R., B.N. McLellan, S. Boutin, D.R. Seip, and S.E. Nielsen. 2011. Developing a population target for an overabundant ungulate for ecosystem restoration. *Journal of Applied Ecology* 48: 935–942.

Sinclair, A.J., L. Sims, and H. Spaling. 2009. Community-based approaches to strategic environmental assessment: Lessons from Costa Rica. *Environmental Impact Assessment Review* 29: 147–156.

Souther, S., M.W. Tingley, V.D. Popescu, D.T.S. Hayman, M.E. Ryan, T.A. Graves, B. Hartl, and K. Terrell. 2014. Biotic impacts of energy development from shale: Research priorities and knowledge gaps. *Frontiers in Ecology and the Environment* 12: 330–338.

Squires, A.J., C.J. Westbrook, and M.G. Dubé. 2010. An approach for assessing cumulative effects in a model river, the Athabasca River Basin. *Integrated Environmental Assessment and Management* 6: 119–134.

Tenenbaum, D.J. 2009. Oil sands development: A health risk worth taking? *Environmental Health Perspectives* 117: 150–156.

Timoney, K., and P. Lee. 2001. Environmental management in resource-rich Alberta, Canada: First world jurisdiction, third world analogue. *Journal of Environmental Management* 63: 387–405.

USGS (US Geological Survey). 2012. Commodity Statistics and Information. http://minerals.usgs.gov/minerals/pubs/commodity/. Accessed 1 Dec 2013.

Wärnbäck, A., and T. Hilding-Rydevik. 2009. Cumulative effects in Swedish EIA practice—difficulties and obstacles. *Environmental Impact Assessment Review* 29: 107–115.

Watkin, L.J., P.S. Kemp, I.D. Williams, and I.A. Harwood. 2011. Managing sustainable development conflicts: The impact of stakeholders in small-scale hydropower schemes. *Environmental Management* 49: 1208–1223.

Wilson, R.R., A.K. Prichard, L.S. Parrett, B.T. Person, G.M. Carroll, M.A. Smith, C.L. Rea, and D.A. Yokel. 2012. Summer resource selection and identification of important habitat prior to industrial development for the Teshekpuk caribou herd in northern Alaska. *PLoS One* 7(11): e48697.

World Bank. 1999. Operational Policy no. 4.01: Environmental Assessment. http://go.worldbank.org/BT7VI5UD50. Accessed 21 Nov 2014.

Zhang, Q., and Z. Lou. 2011. The environmental changes and mitigation actions in the Three Gorges Reservoir region, China. *Environmental Science and Policy* 14: 1132–1138.

Zhu, D., and J. Ru. 2008. Strategic environmental assessment in China: Motivations, politics, and effectiveness. *Journal of Environmental Management* 88: 615–626.

Thompson, [illegible] 2015. [illegible] Alberta, Canada [illegible] [illegible] and [illegible] Journal [illegible]. [illegible] Survey. [illegible] Community [illegible] and [illegible] [illegible]. Accessed [illegible].
[illegible] [illegible] the [illegible] and [illegible] [illegible] 2010. [illegible]
[illegible] Wang [illegible] [illegible] [illegible] [illegible] [illegible] [illegible]
Wang [illegible] [illegible] [illegible] [illegible] [illegible] and [illegible] [illegible] [illegible] [illegible] [illegible] [illegible] [illegible] [illegible] [illegible]
[illegible] [illegible] [illegible] [illegible] [illegible]. Accessed [illegible] 2014.
Zhang [illegible] and [illegible] 2012. The [illegible] change and mitigation [illegible] [illegible] [illegible] [illegible] [illegible] [illegible] [illegible].
Zou, J., and [illegible] 2005. [illegible] and [illegible] and [illegible] [illegible] [illegible] [illegible] [illegible].

# Part II
# Perspectives

# Chapter 3
# Cumulative Impacts and Environmental Values

**Michael P. Gillingham and Chris J. Johnson**

*With contributions by* **Annie L. Booth, Allan B. Costello, Peter L. Jackson, Donald G. Morgan, Justina C. Ray, and Dale Seip**

## 3.1 Introduction

The degradation, conversion, or loss of ecosystem functions is a ubiquitous outcome of human activities and socioeconomic development (MEA 2005). The mechanisms and pathways of change are complex and span many disciplines, including ecology, economics, sociology, policy, planning, and political science. The outcomes of development are often predictable, however, resulting in a global decline in biological diversity as well as in the myriad of other ecosystem services that are the foundation for healthy economies and communities (Perrings et al. 1995; Liu et al. 2003; Smith et al. 2003; Mooney 2010). Even *green* development, such as wind and solar power, has associated environmental costs (Kuvlesky et al. 2007; Pruett et al. 2009), and these changes to natural systems are global, not being restricted to the areas of highest human population density and growth (Johnson et al. 2005; Worm et al. 2006). The alteration of climate will affect even the most pristine ecosystems, and human populations and development opportunities now reach even formerly distant and undisturbed portions of the globe (Thomas et al. 2004).

The cumulative environmental impacts of natural resource development have been well studied (see Krzyzanowski and Almuedo 2010), particularly when compared to the impacts of such activities on human communities, health, and culture. The process of fully including CEA and related impact assessments in project approval, however, continues to mature (see Canter and Ross 2010). Duinker and Greig (2006) provide a brief history of CEA in Canada; they concluded that the process is driven by project approval rather than by the goal of assessing, and potentially limiting, cumulative effects and impacts. The need to consider these

M.P. Gillingham (✉) • C.J. Johnson
The University of Northern British Columbia, Prince George, BC, Canada
e-mail: Michael.Gillingham@unbc.ca; Chris.Johnson@unbc.ca



M.P. Gillingham et al. (eds.), *The Integration Imperative*,
DOI 10.1007/978-3-319-22123-6_3

cumulative effects and to assess their impacts increases as the magnitude and spatial extent of development increases. For example, Northern BC and its associated environments, communities, and Aboriginal peoples, are experiencing an unprecedented level of large-scale industrial projects and smaller-scale supporting work. Such activities are driven by both industry and government, resulting in simultaneous development of multiple overlapping projects, including run-of-river (ROR) electrical generation, coal and gold mining, wind power, transmission lines, hydroelectric dams, oil and natural gas pipelines, and conventional and unconventional exploration and development of petroleum resources.

Over the long term, sustainable development of any environment will depend on a rigorous and defensible assessment of the resultant cumulative impacts (Dubé 2003). In theory, CEA can provide an ongoing method of evaluating impacts and limiting or mitigating projects that exceed the capacity of the local environment. If CEA is restricted to only the development phase of large industrial projects, however, smaller-scale impacts may remain unassessed. Consequently, Duinker and Greig (2006) argued that project-specific CEA is not living up to its potential and is in fact the wrong process for addressing cumulative impacts at regional scales. In this Chapter, we examine a variety of approaches and tools used for assessing and managing cumulative impacts. In so doing, we also set the stage for including sustainable healthy communities into a cumulative impacts framework. We argue that to effectively anticipate and manage for cumulative impacts, we need cross-sector knowledge of past and proposed development, and the ability to plan at spatial scales appropriate to the environmental integrity of the region rather than the project.

## 3.2 The Complexity of Cumulative Effects

Ideally, the assessment of cumulative impacts should focus on a particular area and inform development over time; in practice, assessments are normally triggered by a specific project and occur during the proposal or planning stage. A thorough assessment requires time and the ability to examine a range of scenarios involving many different projects; however, there is an increasing demand for the rapid development of resource extraction projects. Also, cumulative impacts are often best understood when looking back in time. In a landscape that already contains the consequences of socioeconomic development, it can be challenging to separate cumulative from project-specific impacts (Duinker and Greig 2006). As is the case for the impacts on community sustainability and community health, environmental impacts can lead to unexpected consequences. Indirect and unexpected environmental impacts often occur, and may involve extrinsic factors such as climate change (see Vignette 6.3 in Chap. 6). For example, Latham et al. (2011) linked an increase in populations of white-tailed deer in parts of northern Alberta with a large increase in linear features, cut blocks, and roads, and a general increase in young forests associated with large-scale industrial development. This increase in prey led to a dramatic increase in the number of wolves. The combination of the altered landscape and the higher number of predators also led to a dramatic decrease in caribou. As this example illustrates, overlapping cumulative effects often lead to complex and sometimes unexpected and indirect impacts.

Cumulative effects and impacts can cross administrative, jurisdictional, and legislative boundaries. For example, the BC Forest Practices Board (2011) expressed growing concerns about how the cumulative impacts of resource use were affecting the province's land and water, particularly for activities that were not regulated under the *BC Forest and Range Practices Act* (Government of BC 2002b). This concern was based in part on how forest and range resources appear to be afforded different safeguards by government, depending on the industry that is removing or affecting those resources. For example, when forests are managed for forestry, they are managed sustainably, but when the same resource is affected by the development of nonrenewable resources, such as oil and gas, the same sustainability principles do not apply to the affected forest resources.

Other cumulative environmental effects and their resultant impacts occur because of simultaneous demand for the same resource (e.g. water) by many industries. Even with limited or no cumulative impacts on resources, climate change may affect those resources (see Vignette 6.1). Cumulative impacts assessment should, therefore, be an ongoing process that is tied to a functional landscape unit rather than being conducted only as part of an assessment of a specific project. Because any effect, regardless of the footprint of a specific project, can contribute to the cumulative impacts on a landscape, all effects should be given full consideration during impact assessment. Environmental assessment reviews, however, set limits on the size of the project and exempts projects smaller than those limits from consideration, even if the cumulative impact of those projects would be significant. Several projects that affect the same resource and that are being developed at the same time may not be given any consideration in a CEA (see also Box 3.1).

Mitigation and management of potential project-specific environmental impacts is central to CEA. Mitigation may be local or regional and may include limiting rates of extraction, regulating industrial road access, or even defining times or seasons when a given operation is allowed (see Canter and Ross 2010). Mitigation during the life of a project or restoration following completion of the project is often assumed to be an option for nearly all development. However, some outcomes, such as the loss of a species or subspecies, cannot be mitigated. Caribou, for example, are good indicators of intact landscapes, as they are sensitive to human disturbance and industrial development that result in young forests (see Box 3.2). As noted by Chief Roland Willson (West Moberly First Nations):

> So, who is responsible for those caribou, and making sure their habitat is continuous? They need large areas of undisturbed habitat. They are not very industry friendly animals. (Muir and Booth 2012, p. 455)

**Box 3.1 The Growth of Run-of-River Projects in British Columbia (Allan B. Costello)**

The BC Energy Plan (Government of BC 2002c) outlined how the province of BC would become self-sufficient in power generation by 2016, and emphasised the importance of the private sector in new power production (i.e. independent power projects or producers [IPPs]). In particular, the small-scale hydroelectric projects known as ROR projects garnered significant government interest in the decade following the plan's release. Run-of-river projects have been proposed as a greener alternative to traditional hydroelectric facilities because they require small or negligible pondage and divert a portion of the river's flow through penstocks to favourably placed electricity-producing turbines before returning the water to the river downstream. The favourable topography and natural abundance of suitable candidate rivers in BC is conducive to the development of this type of IPP industry.

However, ROR projects are not without environmental consequences. Many such projects substantially alter natural stream hydrographs, particularly when head dams are constructed to regulate stream flows and keep penstock intakes below the water. These head dams often create impassable migration barriers to fish and other aquatic life, and the sediment that accumulates behind such structures can potentially be released downstream in a sudden pulse, with adverse impacts on fish habitat. Diverting large amounts of river water through penstocks further affects water depth and velocity in the section where the diversion occurs, which can often be 3–4 km in length, leading to habitat loss and elevated stream temperatures that are known to adversely affect salmon and other stream fishes. The problem is acute for stream-resident fish species that are isolated above migration barriers in the types of streams that are typically chosen for ROR projects (i.e. streams with a high gradient or with waterfalls). There are also potential cumulative impacts associated with the infrastructure needed to support each project in terms of new roads and stream crossings, clearing of riparian areas to provide room for powerhouses and substations, and transmission lines. These potential impacts have led to debate as to whether sufficient regional planning and impact assessment and sufficient regulatory oversight of RORs is in place, particularly with regard to the cumulative environmental footprint of numerous adjacent projects.

The EA and approval process for ROR projects is generally provincial in scope, with BC Hydro signing electricity purchase agreements with the IPP proponent and the BC Ministry of the Environment issuing water-use licences to legally divert flows and to access and modify public lands. The water-use licence also provides guidance on project design and operational strategies to

(continued)

**Box 3.1** (continued)

minimise impacts on fish, wildlife, and ecosystem values. Given the potential impacts on fish and fish habitat, Fisheries and Oceans Canada often provides further guidance on minimising harm to salmon and other species. Independent power projects that produce less than 50 MW of energy are generally below the EA threshold, exempting many IPPs from this process. Recently, a coalition of conservation groups petitioned the BC Supreme Court to compel the provincial EA office to initiate a full EA for an IPP that involved ten power generation sites along a 40-km stretch of the Holmes River in northeastern BC. Their petition argued that, in combination, those sites would generate more than the 50 MW needed to trigger the provisions of the *BC Environmental Assessment Act* (Government of BC 2002a). The BC Supreme Court ruled, however, that the Holmes River hydroelectric projects should be allowed to proceed without an EA because each of the individual power plants were considered separate projects with a generating capacity of less than 15 MW. Based on this precedent, all proposed ROR projects in BC are evaluated and approved in isolation from other projects, even if they are part of a larger development.

Perhaps one of the largest and most controversial of the proposed ROR projects in BC has been Alterra Power Corporation's Bute Inlet Hydroelectric Project. The Bute Inlet Project proposes the construction of 17 ROR facilities in tributaries in three main watersheds in Bute Inlet on BC's central coast. Although the facilities will have a total peak generating capacity of 1027 MW, only one of the 17 proposed projects (Southgate 1, at 69.5 MW) would have been sufficiently large to trigger the EA process. In this regard, the Bute Inlet Project has been unique; the size of the project, amount of public attention, and likelihood of actions by Fisheries and Oceans Canada and several other federal ministries (including Aboriginal Affairs and Transport Canada) led to the announcement of a joint panel review for the project under the *Canadian Environmental Assessment Act* (Government of Canada 2012). The proponents eventually withdrew their application and the joint review panel was disbanded (CEAA 2012).

Beyond the effects of continued IPPs on individual systems, the more than 700 facilities that have been proposed in BC have raised concerns about the cumulative impacts of multiple ROR projects and the supporting infrastructure. In Bute Inlet alone, 267 km of permanent roads, 142 bridges, a proposed substation, and 443 km of new transmission lines would have been needed to transmit power. In addition to the many terrestrial species at risk from this development, all six species of wild Pacific salmon, steelhead trout, bull trout, rainbow trout, cutthroat trout, Dolly Varden trout, eulachon, and many other fish species occur in the project area; some of these species support a high-quality guided sports fishery. Alterra's own EA documents suggest that potential cumulative impacts associated with this project are expected to be many

(continued)

**Box 3.1** (continued)

and diverse: "Due to the ecologically diverse landscape in the study area, a considerable number of rare ecosystems, plants and animal species could be affected by the construction and operational maintenance of the Project" (Plutonic Hydro Inc. 2008).

The Bute Inlet project is only one of several ROR projects planned for this region. In total, proposed ROR projects for the area would affect 44 rivers from several inlets on this portion of the BC coast. Unfortunately, each of these projects has been largely considered in isolation from any other development, making it impossible to identify the cumulative impacts on ecosystems in the region or to manage them, post-construction, particularly in terms of habitat loss and fragmentation.

**Box 3.2 Cumulative Industrial Impacts on Caribou Herds in the Central Rocky Mountains (Dale Seip)**

As the continental glaciers retreated ca. 14,000 years ago, caribou that had been isolated south of the continental ice sheet began to spread north, and caribou from the Beringia refugium in the north began to spread south, through an ice-free corridor along the eastern slopes of the Rocky Mountains (Weckworth et al. 2012). Today, the woodland caribou that inhabit the eastern slopes of the Rocky Mountains south of the Peace River, BC, display evidence within their mitochondrial DNA of the mixing of these two lineages. For this reason, the caribou within this Central Rocky Mountains population are now considered to be a distinct group (a *Designatable Unit*) within Canada for conservation purposes (COSEWIC 2011).

Caribou continued to live in this area for thousands of years and survived a wide range of environmental conditions and coexisted with natural predators and Aboriginal peoples throughout that period. Within just the past few decades, however, the abundance and distribution of these herds has declined dramatically and they appear to be destined for extinction in the near future (Johnson et al. 2015). The 2013 population estimate for these caribou was only about 615 animals, with about 330 in BC and 286 in Alberta. The Moberly (Klinse-Za) herd has declined from at least 200 in the 1990s to about 16, and the Redrock–Prairie Creek herd has declined from 449 in 1999 to 127 in recent years. The Banff herd and the Burnt-Pine herd have recently been extirpated. There is traditional ecological knowledge (TEK) among Aboriginal peoples as well as other historical evidence that caribou were much more abundant and widespread in the past.

(continued)

**Box 3.2** (continued)

Herds are declining due to a combination of high adult mortality and low calf survival. The annual adult mortality rate ranges from 9–26 % for the different herds. Monitoring of radio-collared caribou has found that predation by wolves is responsible for more than half of the known causes of adult mortality. Calf recruitment rates for the herds range from 9–15 %, which is inadequate to compensate for the adult mortality in many of the herds. The causes of calf mortality have not been studied in the Central Rocky Mountains, but studies in the Northern Rocky Mountains have found that predation by wolves and other predators is the dominant cause (Gustine et al. 2006).

Caribou that winter at high elevations are relatively safe because 97 % of wolf locations in winter occur in low-elevation areas, where they feed on moose, elk, and deer. During the summer, however, wolves make increased use of higher-elevation habitats, where they encounter and kill caribou. Caribou herds that winter in low-elevation forests are in closer proximity to wolves and experience wolf predation year-round.

Although caribou and wolves apparently coexisted in this area for thousands of years, the level of wolf predation has become unsustainable within the past few decades. This period corresponded to widespread industrial development. Forest harvesting, road construction, mining, and oil and gas development altered the low-elevation habitat, whereas mining and gas development occurred in some of the high-elevation habitat. The forest clearing associated with these activities has created areas of early seral shrubland habitats throughout the caribou's range. The recent increase in the amount and distribution of early seral habitat has likely increased the abundance and distribution of moose, elk, and deer and led to an increase in wolf numbers and distribution on caribou ranges. Although caribou are a minor part of the diet of wolves, the level of predation resulting from increased wolf numbers can still be unsustainable for caribou populations. There is some evidence that roads and linear corridors may also enhance the movement of wolves into caribou range (Seip 1992), but the increased numbers of wolves seems to be a more important concern. A similar process is believed to be threatening woodland caribou herds across Canada.

Some industrial development may also directly destroy caribou habitat. Caribou that use forested habitats in winter feed on a combination of arboreal and terrestrial lichens, and removal of those forests will destroy the lichen. Caribou that use high elevations in winter feed primarily on terrestrial lichens that are exposed on windswept slopes. Industrial development such as mining can permanently destroy these habitats.

Activities that disturb and displace caribou from preferred habitats may force them into poorer habitats where the forage quality is lower or the predation

(continued)

**Box 3.2** (continued)

risk is higher. Recently, caribou in the Quintette herd that used to winter in high-elevation ranges have begun to use low-elevation habitats after mining activity expanded into the high elevation winter range. Intensive recreational snowmobiling can also displace caribou from their preferred winter ranges. Such changes in distribution can increase the proximity of caribou to wolves and increase predation-related mortality rates.

Any industrial activity that directly destroys caribou habitat, displaces the caribou, or makes the habitat more suitable for other ungulate prey species can decrease the viability of caribou populations. In much of the province, forest harvesting is the primary industrial activity that threatens caribou habitat. In the Central Rocky Mountains, however, caribou habitat is threatened by the cumulative impacts of forestry, mining, oil and gas extraction, wind power, road construction, and motorised recreation. Those impacts can occur via direct habitat destruction, disturbance, and displacement, or by altering the natural predator–prey system.

Several studies have evaluated trends in the number of caribou or in calf survival relative to the cumulative amount of industrial activity within their range (e.g. Stuart-Smith et al. 1997; James and Stuart-Smith 2000). In these analyses, declining caribou populations or low calf survival were associated with increased amounts of industrial disturbance. Linear disturbances, including roads and seismic lines, are especially problematic because the negative impacts extend over very large areas.

In the Central Rockies, it appears that the level of industrial development that has occurred within the caribou's range over the past few decades has already exceeded a critical threshold and created a situation where caribou populations are no longer sustainable (Johnson et al. 2015). The impact may have been exacerbated by climate change, which has favoured species such as deer and elk. Even if all industrial activities within the caribou's range were to stop immediately, it would take many decades before the habitat recovered. In reality, industrial activities are continuing and expanding into core high-elevation winter range, so the habitat condition continues to deteriorate.

The impact of each individual industrial development usually affects only a fraction of a percent of the overall caribou range, and on its own might be unimportant. However, the cumulative impact of all of the different industrial development activities has resulted in a major alteration of the caribou range, leading to rapidly declining populations.

## 3.3 Issues of Scale and Scaling

Although many types of cumulative impacts can have global consequences, their magnitude and the corresponding responses occur at a range of spatial and temporal scales and across biological and management domains. Thus, processes such as EA that are designed to reduce and mitigate cumulative impacts must fully represent the type and extent of the relevant impacts. In Canada, and for many other jurisdictions, the question of scale is entrenched in the process. During the review of a single project, cumulative impacts are considered in the context of a project footprint or regional study area that would encompass multiple current or reasonably foreseeable future projects. Regional environmental assessment would have a broader geographic and temporal extent, including a requirement for forecasting development activities, and is likely to include a larger number of impacts for consideration than a smaller-scale assessment (Harriman and Noble 2008; Duinker et al. 2012). Regardless of the spatial and temporal scopes, an anthropocentric and reductionist process is used to identify the types and ultimately the number of impacts that are assessed. As was discussed in Sect. 2.2.2, the concept of VECs or analogous terms (e.g. VCs) both focuses the assessment process and constrains it to those components of the environment that are deemed to be important within the context of a project's expected effects.

The project footprint, however, is often defined by the proponent, and this can sometimes result in a misleading CEA. For example, on the Peace River in northern BC, there are currently two hydroelectric dams: the W.A.C. Bennett Dam (completed in 1967) and the Peace Canyon Dam (23 km downriver from the Bennett Dam; completed in 1980). The assessment footprint defined by the BC Hydro and Power Authority for the construction of a third downstream hydroelectric dam on the Peace River (Site C, 83 km downriver from the Peace Canyon dam) did not include the two existing dams. Submissions to the Joint Review Panel by Parks Canada and the governments of Alberta and the NWT questioned the affected area defined by the project, pointing out that the existing hydroelectric development projects have already had downriver impacts on communities and on the Peace-Athabasca Delta (CEAA 2014). Other footprints such as airsheds (i.e. the atmospheric equivalent of watersheds) are often even more difficult to define (see Box 3.3).

The assessment frameworks that are required under EA legislation are tractable from an administrative and even a scientific perspective, but reduce environmental systems to a series of disconnected elements. Broadly defined assessment and management targets, such as maintaining natural levels of biodiversity, ecological function, ecosystem health, or even ecosystem services, are beyond the ability of assessment processes to measure and evaluate relative to a project's impacts (Simberloff 1998; see also Chap. 2). Any review or even a monitoring programme would be overwhelmed by the objective of defining, let alone maintaining, biodiversity or ecosystem health, even though these are laudable objectives that represent functional rather than incomplete and human-relevant elements of complex systems (Wallace 2007).

Since the Millennium Ecosystem Assessment (MEA 2005), the measurement and consideration of ecosystem services has become the new standard for guiding economic and resource development policy and planning at local, national, and international levels (Fisher et al. 2009). Although challenging to quantify, these services can be defined and even monetised according to the value that they provide to human communities both today and potentially in the future (de Groot et al. 2010; see Chap. 5 in relation to human health and well-being). The VECs identified within the EA process are much less holistic than ecosystem services, as they represent only a small subset of the measurable environmental parameters or components that are of value to the human communities that will be affected by a project. There are exceptions to these local-scale impacts. For example, the VECs inherent to the broader value of biodiversity are identified according to provincial or even national assessment processes. In Canada, species listed under the federal *Species at Risk Act* (Government of Canada 2002) will be considered during EA in the context of project-specific and cumulative impacts. Furthermore, the process for recommending species for listing as Threatened or Endangered can consider trends in cumulative habitat loss, but cumulative impacts are not an explicit consideration (see Box 3.4). Once again, however, the conservation of individual species and their habitats represents an incomplete component of biodiversity that is biased by the choice of species to conserve and that lacks any consideration for listing ecosystems or even plant and animal communities that are more holistic representations of biodiversity (Mooers et al. 2007; Findlay et al. 2009; see also Vignette 6.4).

As Richardson (1994) pointed out, the term *value* usually means something of use that is desirable by *Homo sapiens*. For example, values ascribed to wetlands are likely to include habitats for fished or hunted species and services such as wastewater or flood control. From an ecological perspective, those values are rooted in ecological functions found within that wetland or watershed (see Vignette 6.1). In rare cases, the values for a particular landscape may already be prescribed in legislation (e.g. in the *Species at Risk Act*, described in Box 3.4; in the Muskwa–Kechika Management Area, described in Vignette 6.2; see also Shultis and Rutledge 2003). Even then, broad values such as wildlife or wilderness are themselves composed of many different attributes, many of which have competing values.

As we begin to broaden our choice of values beyond those that are traditionally considered as VECs, this will greatly affect the results of a cumulative impact assessment. Therefore, it is important to explicitly recognise that the suite of values used in any cumulative impact assessment is critical. For example, the Government of BC is developing a CEAMF that is designed to enhance the economic and social benefits from resource use while improving environmental outcomes (Government of BC 2014). Their current proposed framework will integrate values such as water and visual quality, key wildlife species, and economic and social well-being into existing business processes and decision making. Similarly, we propose in this book to expand the values considered in cumulative impact assessment to include both community sustainability and health.

Different target values will result in very different impact assessments. Therefore, two of the key initial steps are a full dialogue with stakeholders about the range of values to be considered and an explicit recognition that trade-offs or weighting of competing values will affect the resultant impact assessment. As Johnson and Gillingham (2004) pointed out, using province-wide habitat values to assign a relatively low local value to habitat for a species like grizzly bear does not mean that the habitat is not very important to that species locally. How one decides to make the trade-offs between the potential impacts on wildlife species and the economic value of a development to a community is challenging. Those trade-offs likely should be made at a landscape or regional scale rather than being made a fixed set of values at a provincial scale. Including and/or prioritising values that affect the well-being of people and communities, economics, and ecological systems within a jurisdiction (e.g. Government of BC 2014), however, will influence the results of any assessment.

There is no clear guidance on how to prioritise the choice of VECs or on the spatial and temporal scales at which they should be evaluated. Typically, however, our choice of VECs and the scale at which we quantify the impacts for those VECs is biased by scientific convenience. Johnson and St-Laurent (2011) presented a typology by which to assess the incrementally greater impacts of human activities on biodiversity with application to other environmental values (Fig. 3.1). According to that model, impacts increase as one moves from single obtuse to

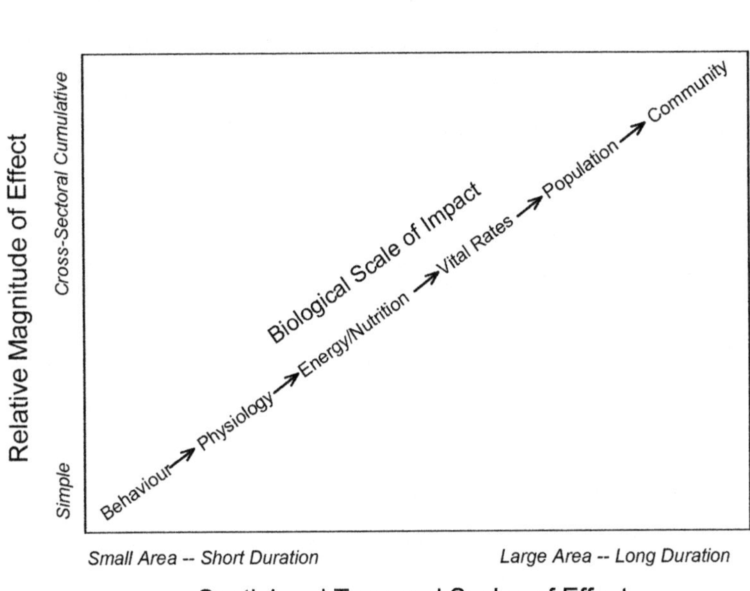


**Fig. 3.1** An illustration of the types of impacts that may occur for wildlife species with increasing magnitude of the effects, with larger scale and longer-term effects, or with a combination of the two. Simple, small-scale, short-term effects may influence only the behaviour of individual animals, perhaps by displacing them from a local area. As the duration and scale of these effects increase and accumulate, the impacts on wildlife species begin to affect their condition (i.e. their physiology, energetics, and nutrition), followed by their vital rates (e.g. recruitment, survival), eventually leading to population-level and even long-term community-level impacts (after Johnson and St-Laurent 2011)

cross-sectoral cumulative impacts that span landscapes and large time periods. In a relative sense, greater levels of impact should be expected in jurisdictions that lack regulations or that fail to effectively implement existing regulations.

Working from the typology of impacts presented by Johnson and St-Laurent (2011), much of our assessment and monitoring science is focused on parameters that can be precisely measured, which are not necessarily the parameters with the most relevance to environmental impacts (Fig. 3.1). When considering broad targets such as biodiversity, dose-response experiments conducted at the scale of the individual animal lead to well-controlled experiments, but the results are not easily extrapolated to population-level processes, including the persistence of species that are at risk. For example, individuals in natural or captive settings can be exposed to some disturbance stimulus and the flight distance, total time of vigilance, or even heart rate can be measured (Bradshaw et al. 1997; Frame et al. 2007; Naylor et al. 2009). Similarly, air and water contaminants can be accurately measured at a point source as units of pollutant (Squires et al. 2010). However, assessing the population- or ecosystem-level impacts of these animal behaviours or pollutants is much more difficult, particularly when considering multiple sources, the interactions among behaviours and pollutants across large spatial and temporal scales, and the confounding effects of other biological or geochemical processes (see Box 3.3).

Fine-scale measures are easy to replicate, more accurate, and, thus, scientifically defensible. Clearly, such measures can assist with understanding and ultimately managing and mitigating human activities. However, given the objectives of EA, which should include such intangibles as sustainability and ultimately the continuance of ecosystem services in all their forms, we should be more concerned with broader impacts on populations and even ecosystems (Duinker and Greig 2006). The effects and resulting impacts have greater relevance as the assessment scales up from the footprint of an individual project to those of multiple projects that result from the activities of several resource sectors that extend across large spatial and temporal scales (see Chap. 7). Mechanistic models can be used to integrate such local-level effects. For example, watershed-scale models can represent the total contributions of pollutants and the potential impacts for human populations (Meulengracht Flachs et al. 2013; see also Vignette 6.1).

**Box 3.3 Cumulative Effects on Air Quality (Peter L. Jackson)**

There are many impacts of outdoor air pollution, ranging from mortality and morbidity of humans, animals, and plants to reduced visibility. These impacts arise from the cumulative effect of elevated concentrations of various pollutants in the ambient (outdoor) air. In regions where there are several sources and types of air pollution, it may be necessary to consider the contribution to ambient pollutant levels from all sources and pollutants together, rather than simply examining each source in isolation, as has been the predominant past practice.

(continued)

**Box 3.3** (continued)

The contributions from multiple sources can be broken down into two components: (a) additive, in which the ambient concentration results from the sum of the concentrations from each individual source, and (b) synergistic, in which the potential addition or removal of pollutants is affected by chemical or other reactions between pollutants or between pollutants and the natural environment.

The concept of an *airshed*—a management unit representing a body of air, by analogy with a watershed—is often used when thinking about the cumulative effects from multiple air pollution sources in a region. This analogy is in some ways poor: in a watershed, gravity and topography constrain the impacts on water to a clearly definable geographic area. In an airshed, however, there is no distinct boundary: the zone of the impact varies in response to changes in meteorology (e.g. wind speed and direction, atmospheric stability and mixing) and terrain. Therefore, cumulative atmospheric impacts are variable in time and space due both to the proximity of the affected area to the sources and due to meteorological conditions and their variable roles in transporting and diffusing atmospheric pollutants.

Although national air quality standards are set cooperatively by the Canadian Council of Ministers of the Environment, whose membership includes the federal, provincial, and territorial Ministers of the Environment, the regulation of air pollutants in Canada is mainly a provincial responsibility; each province and territory has its own regulatory framework for managing air quality. Using BC as an example, the relevant legislation is the *Environmental Management Act* (Government of BC 2003). The Act uses codes of practice, regulations, and site-specific authorizations (both permits and approvals) to manage waste discharge into the atmosphere. Codes of practice and regulations apply across the province and do not allow for a consideration of cumulative effects; provided that industries governed by one of these instruments do not exceed the maximum pollutant amounts, the industries are permitted to operate. In contrast, permits and approvals are site-specific, and allow regional and provincial regulators to consider cumulative impacts before deciding whether to grant permission to emit pollutants into the atmosphere.

BC is also starting to assess cumulative impacts on air quality in a more holistic way. In some regions, voluntary airshed plans are developed by a committee of stakeholders that operate using a consensus-based decision-making approach. These committees attempt to account for all pollution sources and receptors in an airshed. Airshed planning processes have been created in Prince George, Quesnel, and the Bulkley Valley, as well as other regions of BC. It is not clear whether such approaches will be successful in the long term, because they rely on voluntary compliance, and the decision-making process tends to be lengthy. Airshed plans, however, offer the hope of a better-integrated long-term solution that considers the cumulative impacts of multiple sources of pollutant in an area, but at the possible cost of timeliness.

**Box 3.4 Addressing Cumulative Impacts Under Canada's *Species at Risk Act* (Justina C. Ray)**

How an individual species is faring often serves as a valuable indicator (a barometer) of the impacts that arise from human activities. With an increasing number of species being identified as at risk of extinction, laws and policies created by jurisdictions for the explicit purpose of providing protection mechanisms to safeguard these species have been implemented since the 1970s with varying success. The Canadian *Species at Risk Act* (Government of Canada 2002) was enacted in 2002 in order to fulfil Canada's obligations under the United Nations *Convention on Biological Diversity* (CBD 1993). Although multiple tools—including EAs, wildlife management policies and programmes, existing parks and protected areas, stewardship programmes, and regional planning processes—can be employed, each has similar environmental conservation goals. Targeted recovery under the *Species at Risk Act* is designed to uniquely provide an overarching plan into which these other tools should work together to reverse negative trends for a species in Canada.

Under the *Species at Risk Act*, the assessment process to scientifically determine which species are at risk of extinction is separate and distinct from recovery planning and implementation. Threat evaluations can play a critical role in all three processes. Scientific assessments undertaken by the Committee on the Status of Endangered Wildlife in Canada use internationally recognised criteria such as the International Union for Conservation of Nature Red List (IUCN 2014) to classify wildlife species into categories of extinction risk based on species-specific factors that include population size, rate of decline, geographic distribution, and degree of fragmentation of populations (Mace et al. 2008). Description and ranking of threats comprise key sections of both the status report and the recovery strategies—documents that provide the scientific information underlying the assessment and recovery actions, respectively. The relative degree of threat and the likelihood and geographic scope of the threatening events or processes are potentially relevant in almost all quantitative assessment criteria, and are also essential for evaluating those species that belong to the particular category of Special Concern. Likewise, to assist in the recovery of species at risk, factors that are affecting the welfare of individual populations and the species as a whole must be well understood. This means that not only must relevant threats be identified and ranked in terms of their relative impact, but root causes and mechanisms also need to be recognised as much as possible.

Most species face a complex array of threats, many of which interact with one another in a cumulative fashion to elevate the extinction risk. There are, however, a number of aspects of species-at-risk assessment and recovery planning that place limitations on the comprehensive threat assessments that would be required to define the true cumulative impact, and individual threats

(continued)

**Box 3.4** (continued)

tend to be described separately. For example, threat-classification systems tend to focus on direct threats, or "proximate human activities or processes that have impacted, are impacting, or may impact the status of the taxon being assessed" (Master et al. 2012, p. 28). Although the chain of contributing factors that serve as underlying drivers behind any given direct threat should be and often are identified using this framework, threat sections in status reports and recovery strategies have traditionally been limited to a catalogue of stresses with associated narratives.

In spite of the focus on direct threats, the gradual adoption of the objective International Union for Conservation of Nature threat classification scheme (Master et al. 2012) by the Committee on the Status of Endangered Wildlife in Canada in their assessment procedure and in the *Species at Risk Act* recovery processes has led to significant improvements in both assessment and restoration. This system, whereby individual threats are ranked in terms of their scope, severity, and timing, thereby yielding an overall threat score, enables better diagnosis of the most serious plausible threats to survival of a given species using a standardised lexicon. The overall impact on the species is scored based on the number and impact of individual threats (Master et al. 2012). This framework is most helpful for understanding the nature of threatening processes for individual species when: (a) the at-risk status of a species can be attributed to one or more threats that act independently, or (b) the impacts of individual threats are additive or synergistic in nature. However, in the increasingly usual scenario of interacting and cumulative stressors that acted on a species both in the past and that will act on the species into the future, the scheme is less illuminating. Indeed, although the latest (draft) implementation guidance (Environment Canada 2009) by the *Species at Risk Act* urges recovery planners to consider cumulative effects and how individual threats may interact, no instructions are offered for how to do this.

Arguably, species at risk legislation provides an opportunity to embrace a comprehensive treatment of threat abatement, including cumulative impacts, even if in practice this is limited by a perspective that deals with individual species in isolation. Unlike traditional EA processes that are designed to accommodate single proponents and assess individual site-based projects, only the *Species at Risk Act* mandates the gathering in one document of the baseline scientific information for the species' entire Canadian range, and objectives for its overall recovery. The *Species at Risk Act* produces one or more documents that outline a coordinated, inter-jurisdictional plan for action and protection that addresses the socioeconomic and scientific realities in a complete and transparent manner. No other tool identifies range-wide scientific recovery objectives and range-wide critical habitat, thereby combating the complete suite of relevant threats to the species. Unfortunately, when it comes to implementation of restoration actions, the *Species at Risk Act* still relies on other regulatory tools that tend to force a reductionist approach that is inadequate for combating cumulative impacts.

## 3.4 Thresholds and Ecological Theory in Identifying Limits for Cumulative Impacts

In Canada, federal EA legislation requires an assessment of the significance of impacts, both project-specific and cumulative, during the review process. Following the review, the responsible authority or Minister must consider whether the project is likely to cause significant adverse environmental impacts (Government of Canada 2012). These are defined as impacts: that are likely to occur; that represent a relatively large magnitude relative to the persistence, health, quality, or human use of the VEC; and that are difficult to mitigate or manage. Significance increases if the impact occurs continuously or frequently for a long duration over a large area. Thus, spatial and temporal scales are once again important considerations (Fig. 3.1), and are key to effectively and accurately identifying and quantifying cumulative impacts.

Assessments of the significant impacts should be defined with respect to some objective standard. This standard is often considered to represent a *threshold*. The definitions and applications of the threshold concept are founded in resilience theory, and thus have much theoretical and empirical support (May 1977). From an ecological perspective, resilience is the capacity of an ecosystem or community to absorb a disturbance or environmental variation, maintain its basic structure and functions, and avoid transitioning to an alternative stable state (Holling 1973; see Box 3.5). A threshold is the point at which an influential variable controlling ecosystem dynamics is exceeded such that the system suddenly or gradually shifts to a new stable state, perhaps governed by different geochemical or biological processes (Johnson 2013). This can include a change in the overall state of a system from being maintained by natural processes, such as fire or predators, to being controlled by timber or wildlife management.

Unfortunately, thresholds are difficult to detect without extensive research, and are often discovered only after they have been crossed, so establishing them a priori is problematic (Scheffer et al. 2009). The outcomes from a regime shift at any one place in space or time is a new assemblage of species or an entirely different community or ecosystem. Following such a regime shift, feedback mechanisms also change, potentially preventing a natural or even a forced reversal or alteration of the state dynamics to the previous state or a more desired state (Folke et al. 2004). This *hysteresis* effect is a defining element of system resilience and should receive considerable attention when considering the cumulative impacts of development relative to an imperilled or highly valued component of an ecosystem.

Identifying the threshold for an effect or impact is one commonly cited method for assessing its significance, and this approach is thought to be essential for measuring and managing cumulative impacts (Duinker et al. 2012). Thresholds can be related to environmental standards or to guidelines that indicate when the level of some pollutant or activity will substantially affect human health or another VEC such as the conservation status of a protected species. The concept of thresholds has wide application to environmental management and pollution regulation (Brunekreef and Holgate 2002). Laws designed to regulate water or air quality or forest practices are often based on thresholds. This is an intuitive concept and a major component of the business model for industries that rely on the environment for raw materials or as a sink for effluents and emissions. In the process of developing their products, individual companies are held to an assessable (i.e. objective rather than subjective) legally defined standard for pollution, remediation, or retention of some component of the environment. Regulatory thresholds (and related enforcement issues), however, are not without criticism; this is especially the case when the relationship between the regulated activity or pollutant and the environment that it will affect is not well understood (Prüss 1997; Beckers et al. 2002).

Some have suggested that ecological thresholds can be a useful tool for quantifying and then restricting the cumulative impacts of resource development that occurs across landscapes (AXYS Environmental Consulting 2001; Rompré et al. 2010; Environment Canada 2012). Ecological thresholds represent the point at which a non-linear or other substantial change occurs in the dynamics or distribution of an individual organism, population, or community relative to some previous level of disturbance (Ficetola and Denoël 2009). Thus, the threshold can define a stopping point for the approval of additional projects or can suggest when a VEC of interest, such as an endangered species that is limited by the availability of habitat, is reaching the tipping point at which a rapid decline in its population is expected (Amstrup et al. 2010).

The threshold concept is intuitive for land users and natural resource managers, has strong theoretical foundations, has the potential to link directly to regulatory mechanisms, and allows for a consideration of the effects of development both today and in the future. Also, some have suggested or inferred that thresholds offer general guidance that can be applied across multiple populations or species within particular ecosystems (Andrén 1994; Rompré et al. 2010; Environment Canada 2012). Such efficiencies reduce the need for conducting many individual studies and providing location-specific guidance for managing cumulative impacts.

There has been considerable scientific work to identify or attempt to identify thresholds for individual species. At the community level, threshold changes in system functioning and community composition have been well documented. For example, Walker and Meyers (2004) focused on the resilience of plant, animal, and human communities, and presented a database of research that documented and categorised thresholds and shifts in community structure that were related to threshold responses. The majority of the studies listed in the database considered aquatic ecosystems or communities, but the subjects ranged from the persistence or density of single populations to the collapse of ancient civilisations. Contrary to these examples, Schröder et al. (2005) identified and reviewed 35 studies that claimed empirical evidence of alternative community dynamics that resulted from a perturbation that exceeded some threshold. Based on a rigorous set of criteria, they concluded that 13 of the experiments demonstrated the existence of alternative stable states, eight studies conclusively reported no alternative state, and 14 other studies were inconclusive.

Initially, thresholds were proposed as a method for guiding levels of development or constraining impacts in a way that could be generalised within or across ecological systems (Andrén 1994). Many conservation and natural resource practitioners remain hopeful that such targets will meet the needs of a wide range of species, thereby eliminating the requirement for time-consuming and expensive studies focused on individual species (Rompré et al. 2010). Unfortunately, simulation and empirical studies strongly suggest that thresholds, when identified, do not generalise well across species or even across populations (Huggett 2005). For example, Swift and Hannon (2010) summarised the results from 31 different studies across a range of landscape contexts, response variables, and taxa. They reported threshold responses following a change in habitat that encompassed between 1 and 99 % of the pre-disturbance landscape; however, most studies reported thresholds that fell within the range of 10–50 % habitat change.

Even where ecological thresholds are known, they cannot in themselves determine an acceptable level of cumulative impacts (Salmo Consulting 2006). Science-based thresholds, no matter how precise and accurate, only represent information that can assist decision-makers (Kennett 2006). Furthermore, direct socioeconomic values, such as resource development and the associated jobs and tax revenues, are often the priority during land-use and development decision making. As an example, Rompré et al. (2010) and DeLong (2007) reported the *adjustment* of the threshold amount of mature forest thought to maintain natural levels of forest biodiversity across a managed landscape. They found that this threshold was set to a level that facilitated economic development in the form of timber harvesting.

Some who recognise the socioecological realities of landscape management and planning have argued for a more representative definition and application of ecological thresholds. Regulatory limits (Kennett 2006), which represent a decision or management threshold (Martin et al. 2009), have been proposed as a more inclusive representation of the processes and decisions that are necessary for determining stopping points before cumulative impacts become too large. Here, empirical observations, perhaps derived in terms of an ecological threshold, provide guidance relative to the increasing levels of environmental risk, rather than providing precise stopping points for the amount of disturbance that can be allowed before the distribution, abundance, or health of a VEC declines. Regulatory limits are not defined by changes in system drivers or nonlinear tipping points in populations or community dynamics. Instead, the limit is based on the magnitude or extent of anthropogenic disturbance that is permitted before managers anticipate unacceptable ecological change or risk (Johnson 2013). Thus, regulatory limits represent a trade-off between societal values, such as jobs and tax revenues, and ecosystem values, such as the persistence of a species. Some level of risk may be included in this socioecological equation to take advantage of certain ecosystem services at the expense of others. Where the precautionary principle is invoked, the *threshold* may be conservative.

**Box 3.5 Cumulative Impacts and Ecological Resilience (Donald G. Morgan)**

Applying the concept of ecological resilience can help overcome the spatial and temporal limitations of project-based CEA. Ecological resilience—a system's ability to resist transformation to another state or recover from such a transformation—incorporates a systems view of ecosystems, and an understanding of the relationships, nested levels of organisation, and feedbacks among processes within the system and among temporal and spatial scales (Holling 1973). The concepts of ecological resilience provide a framework for understanding the dynamics of a natural resource system and the resulting supply of ecosystem services. Applying an ecological resilience framework in regional CEA shifts the focus from the specific effects of a single project over short periods of time to the broader social–ecological system dynamics and a range of possible futures. Project approval and the potential synergistic impacts of multiple projects can be better evaluated when the system's ecological resilience is considered (Walker et al. 2004).

To ensure the provision of ecosystem services, managing for ecological resilience entails an understanding of the condition and behaviour of coupled human–environment natural resource systems. Ecosystem services are the products of the system that are important to its continuation and its continuing capacity to support people. CEA is also concerned with the management of natural resource systems, but is focused on the maintenance of VECs (Duinker and Greig 2006). Valued ecosystem components are extrinsically valuable; they are the biophysical elements of the natural resource system that people have identified as being important. Separating the condition and behaviour of a system from what these characteristics mean to people is the key to differentiating effects from impacts (see also Vignette 6.5). This separation also aids in understanding the frequent disconnect between science and management in decision making.

Typically, decisions that focus on the allocation of resources, such as water permits and timber supply, are structured around the assumption that a natural resource system will provide an expected amount of a commodity. This approach assumes that the mean of the recent supply of resources from a natural resource system will provide an adequate basis for what to expect in the future. The limitation of this approach is that it does not consider the long-term variability in resource systems, which can be high. British Columbia's mountain pine beetle outbreak, which began in the 1990s, provides an example of how a resource supply can be subject to sudden and extreme variation. Recent management experience with pine-dominated forests did not predict the level of mortality that occurred. In response to this outbreak, timber allocations were increased far beyond what was previously considered sustainable in order to capture some value from the dead and dying pine trees. Historic levels of harvesting in combination with extensive salvage operations have reduced the

(continued)

**Box 3.5** (continued)

availability of future timber that will be necessary to maintain forest-dependent communities (BC MoFLNRO 2012). In contrast, managing for ecological resilience would have considered historical and future forest dynamics more broadly and set annual harvest rates at a lower level. Furthermore, managing for ecological resilience would have anticipated increases in the amount of timber available for harvest that result from large-scale disturbance events (Morgan 2011). Managing for ecological resilience requires: a description of a natural resource system's ecological state; the historical (past), present, and predicted future (e.g. under global warming) disturbances and stressors that have affected, are affecting, and will affect its maintenance; and its possible alternative states, as well as the spatial and temporal scales that are relevant to management (Drever et al. 2006). The current state of a system, such as the structure and composition of a forest, may not be the only state possible, and knowledge of ecological cycles is the key to aligning our expectations of the forest's resource supply with the system's capacity.

Society often decides that the production and management of specific commodities are more desirable than maintaining resource systems in a *wild* state. In many cases, however, such management objectives result in the transformation of a system from a more natural and heterogeneous condition to one that is more homogeneous, structurally different, and controlled by human-dominated processes. Examples include timber and fire management replacing wildfire in the boreal forest, and agriculture replacing fire and bison in tallgrass prairie ecosystems (Scheffer et al. 2001).

Designing and analysing scenarios that describe possible futures is an effective approach for understanding the interplay of impacts from different combinations of projects and natural events (Duinker and Greig 2007). Scenarios make the assumptions about natural resource behaviour explicit and help to clarify which decision-making paths contain the most robust strategies for maintaining ecological resilience across a range of possibilities (Peterson et al. 2003). In some cases, the management objective may not be to maintain ecological resilience, but rather to ensure a deliberate transformation of a system to a new state. A new natural resource system may then emerge that is controlled by a different, human-dominated process. Components of the previous system may no longer be feasible, as in the case of woodland caribou in areas managed for intensive timber production, where their populations will decline (Schaefer 2003; Box 3.2). Evaluating a spectrum of scenarios shifts the focus of assessment from a single project's effects to the collection of forces that act on a natural resource system.

Applying the concepts inherent in ecological resilience—the capacity of an ecosystem to absorb some amount of disturbance without changing its defining structures and processes—greatly benefits CEA. Managing to sustain ecological resilience shifts the focus from individual projects to a systems perspective. This improves our understanding of the collection of stressors that act on ecosystem services and, by association, those aspects of the natural resource system that society values.

## 3.5 Approaches for Assessing Cumulative Impacts

The approaches available for assessing cumulative impacts and the tools available for conducting those assessments are numerous and have been thoroughly reviewed (Canter and Ross 2010; Hegmann and Yarranton 2011; Duinker et al. 2012). Here, we focus on a few tools that are not only important for the assessment of cumulative environmental impacts, but that also require consideration when expanding an assessment framework to account for sustainable communities and community health.

### *3.5.1 Traditional Knowledge*

Almost by definition, cumulative impacts are area-based and such impacts must be considered in the context of the value and use of the affected areas. Aboriginal peoples exemplify this relationship with place. By tradition and occupancy, they are closely linked to the land to obtain food, medicines, and other resources, as well as for their cultural and spiritual well-being. The strong connection between Aboriginal peoples and the land and water that sustains them, both generally and with respect to long-standing use of specific territories, means that some cumulative impacts cannot be mitigated and must instead be prevented. The loss of their culture and of their connection to place cannot be justified by appeals to the greater good of the broader twenty-first century society. In Canada, Aboriginal rights to practice traditional activities and unceded title to their land can and has influenced development decisions that might result in cumulative impacts (see Chap. 2; Boxes 3.6 and 4.3). All too often, however, Aboriginal peoples are forced to resort to legal action to protect important ecosystem components (e.g. Muir and Booth 2012). This contrasts with a more productive and collaborative approach that would involve Aboriginal peoples in a meaningful way in the identification, assessment, and protection of the ecosystems that they value (see Box 5.4; Vignettes 6.6 and 6.7 in Chap. 6; and Bartlett et al. 2012).

Two fundamental limitations of EIAs are a frequent lack of ecological baseline data coupled with an inadequate framework or method to link the ecological and social components of the environment. Aboriginal peoples are not only one of the parties affected by development, they also offer unique perspectives and values that represent the health of the environment. More generally, using long-term and place-specific information in the form of TEK can provide information that otherwise could not be obtained (Azzurro et al. 2011; see Chap. 8). Sallenave (1994) argued that TEK should be formally integrated into the impact assessment process, and Berkes et al. (2000) noted that this knowledge can be of great benefit in assessing cumulative impacts. Heaslip (2008) contrasted the point-by-point assessment of impacts of the coastal aquaculture industry with the more holistic understanding of the broader coastal ecosystem that Aboriginal peoples can provide. However, Usher (2000) pointed out that there are challenges in incorporating both TEK and western science into the review phase of project development.

**Box 3.6 Aboriginal Peoples and Cumulative Impacts (Annie L. Booth)**

Let me tell you a true story. In this story, an Aboriginal man went out hunting one day. He was out hunting not because there wasn't a supermarket down the road (there is), but because hunting is part of who and what he is. Hunting is integral to his culture, spirituality, and history. Hunting is part of his Aboriginal rights, as recognised and protected by the Canadian Constitution and the treaty his Nation signed, so hunting is something the government must protect for this man and his people, at the risk of its honour. As this man is Dunne-za, people who hunt ungulates, he was hunting moose this day (as there are no longer enough caribou to hunt). The man followed the little moose through bush, past an area where lots of moose had bedded down in prime moose habitat. He followed the moose over a ridge and suddenly they were in a clearing with equipment where a gas lease was being developed. The moose kept going through the site and the man followed through some more bush right into a second lease site. The moose went over another ridge and disappeared. The man was told he shouldn't be hunting on land, which was leased to the gas industry. No moose that day, and likely none the next, as beautiful moose habitat disappeared under multiple developments in the bush.

This is how cumulative impacts affect Aboriginal peoples, trying to pursue their culture between ever-increasing numbers of industrial sites, never knowing when critical cultural sites will end up under industrial development. Of never being sure whether there will be a moose to hunt or fish to catch next year or, even if what's caught can be safely eaten.

If non-Aboriginal people could really grasp what it must be like to have your entire culture and sense of self destroyed, inch by inch, they might actually have a start on understanding cumulative impacts. It is not simply a loss of a species, a single cultural practice, or a right. It is multiple, uncoordinated assaults on multiple species, entire ecosystems, and human lives that are made up of interconnections among daily practices, sights, spirituality, cultural realities, meanings of particular landscapes, and when and how you act in all these landscapes, which is connected back to all the species and ecosystems that interact with one's life. It would mean grasping what it means to not be able to hunt caribou or bison anymore because industrial development has affected their populations so deeply and because of how this has affected a culture that has been intimately intertwined with these species for millennia. It might mean understanding why mitigation isn't always appropriate, because sometimes there is no way to mitigate the loss of a practice due to the loss or alteration of an ecosystem component upon which the practice depends. Even if science could restore the ecosystem to resemble its state before development, would the trees and the caribou and the bison truly be the same and could their cultural importance survive? And how can

(continued)

**Box 3.6** (continued)

science mitigate when a man must hunt in the midst of endless industrial development, with new industrial sites springing up one after another, even if one site closes and is *mitigated*?

Understanding the concept of cumulative impacts from an Aboriginal perspective requires an understanding of what might have happened to that moose when it passed over the ridge. Of understanding what happened to all its companions who had bedded down for the night in the trees between gas development sites. Of why caribou and bison are no longer hunted and berries and medicines are picked near roads with trucks roaring past all day and night. Of what it means to take the kids out to learn their culture, only to find the family's campsite destroyed by construction. And of knowing what it means to be afraid to eat what you catch or harvest, or to go home without a moose, not because the moose was luckier than you, but because there are no more places to safely practice a culture and a right that a government swore on its honour to protect and uphold, in exchange perhaps for a few small contracts, a couple of jobs, and some cash to take to that supermarket down the road.

When we understand the layers of meaning and the implications of how our Aboriginal hunter and the moose he was pursuing finding themselves together amidst a gas development project, of what it means for the moose and its kin and for the man and his kin and children, his community, and his culture, we can begin to understand the concept of cumulative impacts. From that understanding, we may find a way to derive a culturally appropriate matrix for cumulative impact assessment and mitigation, and maybe compensation, that also accounts for non-Aboriginal scientific knowledge.

### 3.5.2 Geographic Information Systems: Benefits and Challenges

Proper assessment of cumulative impacts requires support from complex data and information, much of which has a spatial component. Atkinson and Canter (2011) reviewed the many ways that geographic information systems (GIS) can contribute to the assessment of cumulative impacts. In fact, many of the tools used in these assessments have been developed within a GIS framework. Geographic information systems have been used to compare pre-project and post-project scenarios (e.g. Herrmann and Osinski 1999), to map species distributions and habitats (see Rushton et al. 2004), to model non-point sources of both airborne and aquatic pollution (e.g. Chowdary et al. 2004), to explore human health risks (e.g. Bień et al. 2005), to assess visual impacts across viewsheds (e.g. Germino et al. 2001), and to aid in bounding the areas for cumulative impact assessments (Atkinson and Canter 2011). Visual presentation of the cumulative effects on a specific landscape can be very effective in communicating current impacts as well as projected impacts that might occur in the future, and such presentation benefits greatly from access to GIS technology.

Although producing maps of cumulative effects can appear quite simple, maps do not provide a simple method of depicting uncertainty, and each step in mapping cumulative effects requires many assumptions that can increase the uncertainty. geographic information system layers, for example, consist of precise demarcations of points and boundaries (i.e. polygons), and it is difficult to visualise the magnitude of the uncertainty or error associated with these spatial features (see Johnson and Gillingham 2004). Although buffers can be used to demarcate and expand the area of influence or impacts for a particular effect, GIS models often convey a greater sense of precision than is warranted. This is especially the case for modelled or synthetic data, such as maps of wildlife habitat or the distribution of a species (see Johnson and Gillingham 2005). There are specific tools for including social or cultural information such as TEK in a GIS (e.g. Elliot 2008); however, those data layers are often imprecise, and Aboriginal peoples understandably are not willing to simply submit a decontextualized or confidential layer of TEK to support an impact assessment process. Halpern and Fujita (2013) provide a review of many of the assumptions involved in mapping cumulative effects, including the relative importance and accuracy of mapping different stressors, and how ecosystem changes can differ in response to the same individual stressors.

### 3.5.3 Scenario Planning Approaches

The future is never known, but scenario-based planning provides a way to identify and explore the uncertainties and driving forces in an environment. However, scenario planning is not about *predicting* the future, but rather about *exploring* possible futures. As an example, the Government of Alberta is currently undertaking a regional strategic assessment of the South Athabasca Oil Sands area. Scenario planning is being used to characterise the potential cumulative impacts of various levels of development resulting from future oil sands operations (Government of Alberta 2009).

In its current form, CEA is often reactive and considers the effects of new development projects or ongoing changes and their impacts on the environment (see the review by Duinker et al. 2012). Past, present, and future projects must be considered if cumulative impacts are to be quantified fully (Greig et al. 2004). The uncertainty in predicting the future or documenting the past can be addressed by constructing and comparing a number of scenarios that represent plausible development activities and the resulting impacts. Those scenarios could be developed by technical specialists (e.g. Strimbu and Innes 2012), by stakeholders and communities (e.g. GSHSAC 2007), or (as we are proposing throughout this book) by a combination of both groups. Duinker et al. (2012) suggest that practitioners typically develop highly constrained scenarios based on past and future effects. If the goal is to meaningfully engage a community and influence a decision-making process, then public and stakeholder involvement in scenario development are critical.

Various tools are available to represent and quantify the cumulative impacts resulting from development scenarios. ALCES, A Landscape Cumulative Effects Simulator (ALCES Landscape and Land-Use Ltd., www.alces.ca) can be used to track overlapping land uses and natural disturbance regimes in terrestrial and aquatic landscapes. The ALCES simulator can report on a full complement of economic, ecological, and social indicators. For example, Schneider et al. (2003) used ALCES to build a landscape-scale simulation model to forecast the cumulative impacts of present and future forestry and energy development in northern Alberta. They modelled a series of *best practices* and found that much better ecological outcomes could be achieved while still maintaining economic benefits. Another approach was demonstrated by Noble (2008), who used MARXAN (Ball et al. 2009) to conduct a regional CEA in the Great Sand Hills region of Saskatchewan. MARXAN is decision-support software that uses optimization algorithms (see Ball 2000) to accommodate perturbations to habitats while attempting to meet biodiversity targets; to do so, it assesses landscape disturbance under various scenarios.

## 3.6 Looking Forward: Cumulative Impacts and the Role of Land-Use Planning

A number of approaches for assessing and limiting cumulative impacts have been tested, and transcend the many limitations of project-specific assessment and approval (see Chap. 2; Duinker and Greig 2006). When compared with the restricted scope and reactive nature of EA legislation, RSEA provides broader and more forward-looking, decision-making opportunities (Gunn and Noble 2009). A complementary approach, strategic land-use planning, has the potential to integrate many of the proven and successful elements of CEAMFs and RSEA. Furthermore, there is a formal recognition of the benefits of land-use planning and much experience in the application of planning principles to conservation and the management of natural resources across large regions of western Canada (Haddock 2010). There are parallel planning approaches for economic development, but these place much less emphasis on the interactions between the development of natural resources and the functioning of environmental systems (Roberts et al. 2006).

Strategic land-use planning is designed to develop goals and objectives for a range of land uses across a region, followed by strategies to achieve these objectives, and finally monitoring to ensure that the goals and objectives are met. In addition to maintaining specific environmental values and sustainable resource development that supports dependent communities, such exercises could be oriented to develop objectives and targets that explicitly regulate cumulative impacts. Strategic planning differs from operational management processes in that the planning landscape and goals are broader and typically involve a number of resource sectors and a wider range of economic and environmental values. Although past planning efforts in BC were focused on forestry, the diversification and growth of the province's resource economy has required the consideration of a larger number of economic values while still recognising the goal of environmental sustainability. Given the diversity of values and participants at a planning table, government involvement is essential to coordinate the process, demand participation, and regulate the outcomes in the form of development objectives. The pathways between land-use planning and EA are well established, especially in the context of cumulative impacts (e.g. Chap. 2; Noble 2009). As Bardecki (1990, p. 322) noted, "Assessing and managing cumulative impacts is planning".

In North America, BC is arguably the most experienced jurisdiction in implementing regional, comprehensive, and participatory land-use planning (Frame et al. 2004). This experience originates from a set of provincially driven planning exercises that were initiated in the early 1990s. Of particular note, Land and Resource Management Planning (LRMP) was a consensus-based sub-regional land-use model that was implemented for the majority of the province (Jackson and Curry 2002). Across the Muskwa-Kechika Management Area (MKMA) in northeastern BC, the government developed a more focused planning process. Before allocating rights to oil and gas reserves, large planning units are subjected to a pre-tenure planning process. Similar in scope to the LRMP process, various stakeholder, community, and Aboriginal groups identify important ecological, cultural, and socioeconomic values across the planning area and then work collaboratively to define objectives, strategies, and indicators to ensure that oil and gas activities do not negatively affect those values (see Vignette 6.2). Aside from limited monitoring, however, there is no procedural or legislated recognition of the cumulative impacts of those activities during either the exploration or the development phase.

The LRMP process in BC is now much reduced, as more than 85 % of the public land base (a total of 97 million ha) is now covered by 26 land-use plans (ILMB 2006). The process cost from $50–$100 million since its inception, with the cost per plan escalating as a result of more meaningful engagement with Aboriginal peoples (ILMB 2006). Looking forward, however, there is substantial legacy in the form of human capital and process that could be applied to the strategic management of cumulative impacts. The province developed an evolving process for engaging local communities and stakeholder groups to articulate a vision for the intensity, type, and location of land-use activities. Bringing the planning process to the affected communities allowed the public to gain a much better understanding of the complexity of resource management; it also developed local-level relationships and trust between interest groups, the resource industry, and government (Frame et al. 2004). The consensus-based decision-making process was successful in engaging a wide range of interest groups in dialogue and decision making that often involved difficult and controversial trade-off decisions (Mascarenhas and Scarce 2004). Although not without limitations and faults, the planning process had an unprecedented scope in terms of both geography and the diversity of values and communities involved.

The strategic land-use planning process and many of its outcomes in BC meet the goals of large-scale and inclusive assessment and management of cumulative impacts (Gunn 2009; Noble 2009). Alex Grzybowski and Associates (2001) reviewed the BC LRMP process and reported a strong correspondence with the characteristics of RSEA. Both processes: operate at large spatial and temporal scales; are comprehensive, as they consider a broad range of ecological and socio-economic effects, often across interacting resource sectors; assume a proactive and strategic perspective; demand collaboration and the involvement of multiple agencies, stakeholder groups, and Aboriginal peoples; are time-consuming, as they take several years to complete; and require effectiveness monitoring and adaptation to meet changing socioeconomic priorities and environmental conditions. They noted that the most important outcome of strategic land-use plans was the articulation of a strategic vision by stakeholders at the planning tables.

Notwithstanding this progress, there is still considerable opportunity to plot a future for many of the country's developing landscapes (Timoney and Lee 2001). Yet, if past experience is a guide, there is also a high likelihood of future conflict between resource development and environmental and community values. Previous experience suggests that some process for developing a strategic vision will be essential for success. By definition, cumulative impacts are the anticipated, unanticipated, planned, and unplanned outcomes of piecemeal development. Strategic planning would allow communities to work with governments and industry to predict and plan for the type and magnitude of the impacts that will occur across multiple landscapes. These are sensible and progressive objectives that are not facilitated by current EA legislation.

## 3.7 Conclusions

Although cumulative impacts are often best understood by looking back in time, we increasingly need to anticipate future impacts while new development projects are being planned. This follows from the philosophy that it is generally easier to prevent an adverse impact than it is to mitigate that impact or restore the original system state. Therefore, the assessment of cumulative impacts must be an ongoing process, not simply a one-time step in the project approval process. Although appropriate spatial scales are easier to define for some projects than others (e.g. watersheds versus airsheds), cumulative impacts must always be assessed at appropriate spatial and temporal scales, and are best tied to a functional landscape unit rather than being conducted only as part of an assessment for a specific project. Consequently, when several projects affect the same resource, they must be considered together within a cumulative effects framework, whether they will be developed sequentially or at the same time.

The concepts of mitigation and management of project-specific impacts are central to CEA, but we must recognise that many of these impacts, such as the loss of a species or subspecies, cannot be mitigated and must instead be prevented. At the same time, the conservation of individual species does not adequately represent the other elements of biodiversity. There is bias in the choice of species that should be conserved and there is little consideration for how that bias will affect the functioning of ecosystems or even plant and animal communities that include that species. It is important to recognise that there are trade-offs in selecting the target values for the CEA; using different target values will result in very different impact assessments. Economics, as an affected value, is part of, not separate from, sustainability, as sustainability across all values is the goal.

Recognition of the connection between economics and environmental sustainability serves to introduce and highlight the need for a greater consideration of the community impacts that may arise as a function of or independent of the environmental changes that are typically assessed as cumulative impacts (see Chaps. 2, 4 and 5). The range of community impacts is broad, as these include the social, cultural, and economic changes caused by rapid and cumulative development activities. Linking the environment with the community creates a conceptual space in which health impacts can be considered, both in terms of ecosystem health and in terms of community and individual health. Greater attention to community and health impacts—the subject of the following chapters—will also mean accounting for and expanding the same critical issues that have been raised in critiques of environmental impacts: it is necessary to define appropriate spatial and temporal scales, clarify the values and trade-offs, and pay attention to legacy impacts that began before any new assessment.

Just as EA processes were historically broadened to represent a greater range of perspectives, increased attention to community and health impacts will mean greater attention to public participation. Perhaps most importantly in this regard, Aboriginal peoples must be considered as equal partners in development planning and in prevention, mitigation, or management of the resulting cumulative impacts. Aboriginal peoples bring a unique cultural perspective and formally recognised legal and moral rights to all discussions of the future, particularly in the form of TEK that can complement scientific approaches and further inform past and potential impacts. The key to successfully minimising cumulative impacts before they require mitigation is pre-planning. To be successful, this pre-planning needs to be inclusive, must take place at a landscape or larger scale, and must be able to account for and affect development across sectors.

Some form of regulatory limits, provided they are derived transparently, based in ecology, and account for trade-offs among values, will be essential. Limits may not be politically convenient, as they imply some change in the rate or type of economic activity, but they are necessary for finding a balance between economic, social, and environmental values. Tools such as scenario planning and GIS are useful for informing broader (i.e. community and health) and more strategic efforts to address cumulative impacts. Typically, however, these tools do not adequately accommodate or portray uncertainty or errors in prediction. A range of scenarios based on different assumptions must be examined, constrained by the recognition that there is no single correct future that represents all of the positive and negative impacts of development.

## References

Alex Grzybowski and Associates. 2001. *Regional environmental effects assessment and strategic land use planning in British Columbia*. Ottawa: Canadian Environmental Assessment Agency Research and Development Monograph Series.

Amstrup, S.C., E.T. DeWeaver, D.C. Douglas, B.G. Marcot, G.M. Durner, C.M. Bitz, and D.A. Bailey. 2010. Greenhouse gas mitigation can reduce sea-ice loss and increase polar bear persistence. *Nature* 468: 955–958.

Andrén, H. 1994. Effects of habitat fragmentation on birds and mammals in landscapes with different proportions of suitable habitat: A review. *Oikos* 71: 355–366.

Atkinson, S.F., and L.W. Canter. 2011. Assessing the cumulative effects of projects using geographic information systems. *Environmental Impact Assessment Review* 31: 457–464.

AXYS Environmental Consulting. 2001. *Thresholds for addressing cumulative effects on terrestrial and avian wildlife in the Yukon*. Whitehorse: Department of Indian and Northern Affairs and Environment Canada.

Azzurro, E., P. Moschella, and F. Maynou. 2011. Tracking signals of change in Mediterranean fish diversity based on local ecological knowledge. *PLoS One* 6(9): e24885.

Ball, I.R. 2000. Mathematical applications for conservation ecology: The dynamics of tree hollows and the design of nature reserves. PhD Thesis, University of Adelaide.

Ball, I.R., H.P. Possingham, and M. Watts. 2009. Marxan and relatives: Software for spatial conservation prioritisation. In *Spatial conservation prioritisation: Quantitative methods and computational tools*, ed. A. Moilanen, K.A. Wilson, and H.P. Possingham, 185–195. Oxford: Oxford University Press.

Bardecki, M.J. 1990. Coping with cumulative impacts: An assessment of legislative and administrative mechanisms. *Impact Assessment Bulletin* 8: 319–344.

Bartlett, C., M. Marshall, and A. Marshall. 2012. Two-eyed seeing and other lessons learned within a co-learning journey of bringing together Indigenous and mainstream knowledges and ways of knowing. *Journal of Environmental Studies and Sciences* 2: 331–340.

BC Forest Practices Board. 2011. *Cumulative effects: From assessment towards management*. Victoria: Forest Practices Board. Special Report 39.

BC MoFLNRO (British Columbia Ministry of Forests Lands and Natural Resource Operations. 2012. Mid-term timber supply: Lakes Timber Supply Area. Victoria: BC Government. http://www.for.gov.bc.ca/hfp/mountain_pine_beetle/mid-term-timber-supply-project/Lakes%20TSA.pdf. Accessed 14 Nov 2014.

Beckers, J., Y. Alila, and A. Mtiraoui. 2002. On the validity of the British Columbia Forest Practices Code guidelines for stream culvert discharge design. *Canadian Journal of Forest Research* 32: 684–692.

Berkes, F., J. Colding, and F. Carl. 2000. Rediscovery of traditional ecological knowledge as adaptive management. *Ecological Applications* 10: 1251–1262.

Bień, J.D., J. ter Meer, W.H. Rulkens, and H.H.M. Rijnaarts. 2005. A GIS-based approach for the long-term prediction of human health risks at contaminated sites. *Environmental Modeling & Assessment* 9: 221–226.

Bradshaw, C.J.A., S. Boutin, and D.M. Hebert. 1997. Effects of petroleum exploration on woodland caribou in northeastern Alberta. *Journal of Wildlife Management* 61: 1127–1133.

Brunekreef, B., and S.T. Holgate. 2002. Air pollution and health. *The Lancet* 360: 1233–1242.

Canter, L., and B. Ross. 2010. State of practice of cumulative effects assessment and management: The good, the bad and the ugly. *Impact Assessment and Project Appraisal* 28: 261–268.

CBD (Convention on Biological Diversity). 1993. United Nations Convention on Biological Diversity. http://www.cbd.int/convention/text/. Accessed 20 Jan 2015.

CEAA (Canadian Environmental Assessment Agency). 2012. Bute Inlet Hydroelectric Project: Environmental Assessment Terminated. https://www.ceaa-acee.gc.ca/default.asp?lang=En&xml=FC11690D-7C93-441E-9F72-9EF48AF1A7D9. Accessed 11 Nov 2014.

———. 2014. *Report of the Joint Review Panel: Site C Clean Energy Project, BC Hydro and Power Authority, British Columbia*. Beaconsfield: Canadian Environmental Assessment Agency.

Chowdary, V.M., S. Kar Yatindranath, and S. Adiga. 2004. Modelling of non-point source pollution in a watershed using remote sensing and GIS. *Journal of the Indian Society of Remote Sensing* 32: 59–73.

COSEWIC. 2011. *Designatable Units for Caribou (Rangifer tarandus) in Canada*. Ottawa: Committee on the Status of Endangered Wildlife in Canada.

de Groot, R.S., R. Alkemade, L. Braat, L. Hein, and L. Willemen. 2010. Challenges in integrating the concept of ecosystem services and values in landscape planning, management, and decision making. *Ecological Complexity* 7: 260–272.

DeLong, S.C. 2007. Implementation of natural disturbance-based management in northern British Columbia. *Forestry Chronicle* 83: 338–346.

Drever, C.R., G.D. Peterson, C. Messier, Y. Bergeron, and M. Flannigan. 2006. Can forest management based on natural disturbance maintain ecological resilience? *Canadian Journal of Forest Research* 36: 2285–2299.

Dubé, M.G. 2003. Cumulative effect assessment in Canada: A regional framework for aquatic ecosystems. *Environmental Impact Assessment Review* 23: 723–745.

Duinker, P.N., and L.A. Greig. 2006. The impotence of cumulative effects assessment in Canada: Ailments and ideas for redeployment. *Environmental Management* 37: 153–161.

———. 2007. Scenario analysis in environmental impact assessment: Improving explorations of the future. *Environmental Impact Assessment Review* 27: 206–219.

Duinker, P.N., E.L. Burbidge, S.R. Boardley, and L.A. Greig. 2012. Scientific dimensions of cumulative effects assessment: Toward improvements in guidance for practice. *Environmental Review* 21: 40–52.

Elliot, N.J. 2008. Including Aboriginal values in resource management through enhanced geospatial communication. PhD Dissertation, University of Northern British Columbia.

Environment Canada. 2009. *Species at Risk Act Policies: Overarching Policy Framework. Draft Document*. Ottawa: Government of Canada. www.sararegistry.gc.ca/document/com982/tdm-toc_e.cfm. Accessed 17 Nov 2014.

———. 2012. *Recovery Strategy for the Woodland Caribou (Rangifer tarandus caribou), Boreal Population, in Canada*, Species at risk act recovery strategy series. Ottawa: Environment Canada.

Ficetola, G.F., and M. Denoël. 2009. Ecological thresholds: An assessment of methods to identify abrupt changes in species-habitat relationships. *Ecography* 32: 1075–1084.

Findlay, C.S., S. Elgie, B. Giles, and L. Burr. 2009. Species listing under Canada's Species at Risk Act. *Conservation Biology* 23: 1609–1617.

Fisher, B., R.K. Turner, and P. Morling. 2009. Defining and classifying ecosystem services for decision making. *Ecological Economics* 68: 643–653.

Folke, C., S. Carpenter, B. Walker, M. Scheffer, T. Elmqvist, L. Gunderson, and C.S. Holling. 2004. Regime shifts, resilience, and biodiversity in ecosystem management. *Annual Review of Ecology, Evolution, and Systematics* 35: 557–581.

Frame, T.M., T. Gunton, and J.C. Day. 2004. The role of collaboration in environmental management: An evaluation of land and resource planning in British Columbia. *Journal of Environmental Planning and Management* 47: 59–82.

Frame, P.F., H.D. Cluff, and D.S. Hik. 2007. Response of wolves to experimental disturbance at homesites. *Journal of Wildlife Management* 71: 316–320.

Germino, M.J., W.A. Reiners, B.J. Blasko, D. McLeod, and C.T. Bastian. 2001. Estimating visual properties of Rocky Mountain landscapes using GIS. *Landscape and Urban Planning* 53: 71–83.

Government of Alberta. 2009. *Terms of reference for developing the Lower Athabasca Regional Plan*. Edmonton: Government of Alberta. www.ceaa.gc.ca/050/documents/45582/45582E.pdf. Accessed 20 Jan 2015.

Government of BC. 2002a. BC Environmental Assessment Act. (S.BC. 2002. C. 43). http://www.bclaws.ca/Recon/document/ID/freeside/00_02043_01. Accessed 29 Mar 2015.

———. 2002b. BC Forest and Range Practices Act. (S.BC. 2002, C. 69). http://www.bclaws.ca/Recon/document/ID/freeside/00_02069_01. Accessed 29 Mar 2015.
———. 2002c. *Energy for our future: A plan for BC*. Vancouver: BC Ministry of Energy, Mines and Petroleum Resources. Retrieved from http://tinyurl.com/kba83kz. Accessed 11 November 2014.
———. 2003. BC Environmental Management Act. (S.BC. 2003. C. 53). http://www.bclaws.ca/Recon/document/ID/freeside/03053_00, Accessed 29 Mar 2015.
———. 2014. *Addressing cumulative effects in natural resource decision-making: A framework for success*. Victoria: Government of BC. Cumulative Effects Framework Overview Report. http://www2.gov.bc.ca/gov/DownloadAsset?assetId=F2A8B8AE894348DBA4CF7942EC592762. Accessed 1 Apr 2015.
Government of Canada. 2002. Species at Risk Act, 2002 (S.C. 2002, C. 29). http://laws-lois.justice.gc.ca/eng/acts/S-15.3/. Accessed 29 Mar 2015.
———. 2012. Canadian Environmental Assessment Act, 2012 (S.C. 2012, C. 19, S. 52). http://laws-lois.justice.gc.ca/eng/acts/c-15.21/page-1.html. Accessed 12 Dec 2014.
Greig, L., K. Pawley, and P. Duinker. 2004. *Alternative scenarios of future development: An aid to cumulative effects assessment*. Gatineau: Environmental Assessment Agency.
GSHSAC (Great Sand Hills Scientific Advisory Committee). 2007. *Great sand hills regional environmental study*. Regina: Canada Plains Research Center. http://www.environment.gov.sk.ca/GreatSandHillsRegionalEnvironmentalStudy. Accessed 1 Nov 2014.
Gunn, J.H. 2009. Integrating strategic environmental assessment and cumulative effects assessment in Canada. PhD Dissertation, University of Saskatchewan.
Gunn, J.H., and B.F. Noble. 2009. A conceptual basis and methodological framework for regional strategic environmental assessment (R-SEA). *Impact Assessment and Project Appraisal* 27: 258–270.
Gustine, D.D., K.L. Parker, R.J. Lay, M.P. Gillingham, and D.C. Heard. 2006. Calf survival of woodland caribou in a multi-predator ecosystem. *Wildlife Monographs* 165: 1–32.
Haddock, M. 2010. *Environmental assessment in British Columbia*. Victoria: Environmental Law Centre, University of Victoria.
Halpern, B.S., and R. Fujita. 2013. Assumptions, challenges, and future directions in cumulative impact analysis. *Ecosphere* 4(10): 131.
Harriman, J.A., and B.F. Noble. 2008. Characterizing project and strategic approaches to regional cumulative effects assessment in Canada. *Journal of Environmental Assessment Policy and Management* 10: 25–50.
Heaslip, R. 2008. Monitoring salmon aquaculture waste: The contribution of First Nations' rights, knowledge, and practices in British Columbia, Canada. *Marine Policy* 32: 988–996.
Hegmann, G., and G.A. Yarranton. 2011. Alchemy to reason: Effective use of cumulative effects assessment in resource management. *Environmental Impact Assessment Review* 31: 484–490.
Herrmann, S., and E. Osinski. 1999. Planning sustainable land use in rural areas at different spatial levels using GIS and modelling tools. *Landscape and Urban Planning* 46: 93–101.
Holling, C. 1973. Resilience and stability of ecological systems. *Annual Review of Ecology, Evolution, and Systematics* 4: 1–23.
Huggett, A.J. 2005. The concept and utility of 'ecological thresholds' in biodiversity conservation. *Biological Conservation* 124: 301–310.
ILMB (Integrated Land Management Bureau). 2006. *A new direction for land use planning in BC*. Victoria: Integrated Land Management Bureau, Ministry of Agriculture and Lands.
IUCN (International Union for Conservation of Nature). 2014. IUCN Red List of Threatened Species. Version 2014.2. www.iucnredlist.org. Accessed 20 Jan 2015.
Jackson, T., and J. Curry. 2002. Regional development and land use planning in rural British Columbia: Peace in the woods? *Regional Studies* 36: 439–443.
James, A.R.C., and A.K. Stuart-Smith. 2000. Distribution of caribou and wolves in relation to linear corridors. *Journal of Wildlife Management* 64: 154–159.
Johnson, C.J. 2013. Identifying ecological thresholds for regulating human activity: Effective conservation or wishful thinking? *Biological Conservation* 168: 57–65.

Johnson, C.J., and M.P. Gillingham. 2004. Mapping uncertainty: Sensitivity of wildlife habitat ratings to expert opinion. *Journal of Applied Ecology* 41: 1032–1042.

———. 2005. An evaluation of mapped species distribution models used for conservation planning. *Environmental Conservation* 32: 117–128.

Johnson, C.J., and M.-H. St-Laurent. 2011. Unifying framework for understanding impacts of human developments on wildlife. In *Energy development and wildlife conservation in Western North America*, ed. D.E. Naugle, 23–54. Washington: Island Press.

Johnson, C.J., M.S. Boyce, R.L. Case, H.D. Cluff, R.J. Gau, A. Gunn, and R. Mulders. 2005. Quantifying the cumulative effects of human developments: A regional environmental assessment for sensitive Arctic wildlife. *Wildlife Monographs* 160: 1–36.

Johnson, C.J., L.P.W. Ehlers, and D.R. Seip. 2015. Witnessing extinction: Cumulative impacts across landscapes and the future loss of an evolutionarily significant unit of woodland caribou in Canada. *Biological Conservation* 186: 176–186.

Kennett, S. 2006. *From science-based thresholds to regulatory limits: Implementation issues for cumulative effects management*. Yellowknife: Environment Canada.

Krzyzanowski, J., and P.L. Almuedo. 2010. *Cumulative impacts of natural resource development on ecosystems and wildlife: An annotated bibliography for British Columbia*. Prepared for the BC Ministry of Forests and Range, Forest Investment Account-Forest Science Program. Kamloops: Forrex Forum for Research and Extension in Natural Resources.

Kuvlesky, W.P., L.A. Brennan, M.L. Morrison, K.K. Boydston, B.M. Ballard, and F.C. Bryant. 2007. Wind energy development and wildlife conservation: Challenges and opportunities. *Journal of Wildlife Management* 71: 2487–2498.

Latham, A.D.M., M.C. Latham, N.A. McCutchen, and S. Boutin. 2011. Invading white-tailed deer change wolf–caribou dynamics in northeastern Alberta. *Journal of Wildlife Management* 75: 204–212.

Liu, J., G. Daily, P.R. Ehrlich, and G.W. Luck. 2003. Effects of household dynamics on resource consumption and biodiversity. *Nature* 421: 530–533.

Mace, G.M., N.J. Collar, K.J. Gaston, C. Hilton-Taylor, H.R. Akcakaya, N. Leader-Williams, E.J. Milner-Gulland, and N. Stuart. 2008. Quantification of extinction risk: IUCN's system for classifying threatened species. *Conservation Biology* 22: 1424–1442.

Martin, J., M.C. Runge, J.D. Nichols, B.C. Lubow, and W.L. Kendall. 2009. Structured decision making as a conceptual framework to identify thresholds for conservation and management. *Ecological Applications* 19: 1079–1090.

Mascarenhas, M., and R. Scarce. 2004. "The intention was good": Legitimacy, consensus-based decision making, and the case of forest planning in British Columbia, Canada. *Society and Natural Resources* 17: 17–38.

Master, L.L., D. Faber-Langendoen, R. Bittman, G.A. Hammerson, B. Heidel, L. Ramsay, K. Snow, A. Teucher, and A. Tomaino. 2012. *NatureServe conservation status assessments: Factors for Evaluating Species and Ecosystem Risk*. Arlington: NatureServe.

May, R.M. 1977. Thresholds and breakpoints in ecosystems with a multiplicity of stable states. *Nature* 269: 471–477.

MEA (Millennium Ecosystem Assessment). 2005. *Ecosystems and human well-being: Synthesis*. Washington: Island Press.

Meulengracht Flachs, E., J. Sørensen, J. Bønløkke, and H. Brønnum-Hansen. 2013. Population dynamics and air pollution: The impact of demographics on health impact assessment of air pollution. *Journal of Environmental and Public Health* 2013: 760259.

Mooers, A.Ø., L.R. Prugh, M. Festa-Bianchet, and J.A. Hutchings. 2007. Biases in legal listing under Canadian endangered species legislation. *Conservation Biology* 21: 572–575.

Mooney, H.A. 2010. The ecosystem-service chain and the biological diversity crisis. *Philosophical Transactions of the Royal Society B* 365: 31–39.

Morgan, D.G. 2011. Exploring the social–ecological resilience of forest ecosystem services. MSc Thesis, University of Northern British Columbia.

Muir, B., and A. Booth. 2012. An environmental justice analysis of caribou recovery planning, protection of an Indigenous culture, and coal mining development in northeast British Columbia, Canada. *Environment, Development and Sustainability* 14: 455–476.

Naylor, L.M., M.J. Wisdom, and R.G. Anthony. 2009. Behavioral responses of North American elk to recreational activity. *Journal of Wildlife Management* 73: 328–338.

Noble, B. 2008. Strategic approaches to regional cumulative effects assessment: A case study of the Great Sand Hills, Canada. *Impact Assessment and Project Appraisal* 26: 78–90.

Noble, B.F. 2009. Promise and dismay: The state of strategic environmental assessment systems and practices in Canada. *Environmental Impact Assessment Review* 29: 66–75.

Perrings, C., K.G. Maler, C. Folke, C.S. Holling, B.O. Jansson, M. Weitzman, D.W. Schindler, B.W. Walker, J. Roughgarden, R. Constanza, M. Kemp, W. Boynton, R.K. Turner, I.M. Gren, I. Bateman, G. Brown, C. Perrings, T. Swanson, E.B. Barbier, M. Rauscher, and S. Barrett. 1995. *Biodiversity loss: Economic and ecological issues*. Cambridge: Cambridge University Press.

Peterson, G.D., G.S. Cumming, and S.R. Carpenter. 2003. Scenario planning: A tool for conservation in an uncertain world. *Conservation Biology* 17: 358–366.

Plutonic Hydro Inc. 2008. Revised Project Description for the Bute Inlet Hydroelectric Project, Dec. 19, 2008. www.ceaa-acee.gc.ca/050/documents/34284/34284E.pdf. Accessed 11 Jan 2015.

Pruett, C.L., M.A. Patten, and D.H. Wolfe. 2009. It's not easy being green: Wind energy and a declining grassland bird. *Bioscience* 59: 257–262.

Prüss, A. 1997. Review of epidemiological studies on health effects from exposure to recreational water. *International Journal of Epidemiology* 27: 1–9.

Richardson, C. 1994. Ecological functions and human values in wetlands: A framework for assessing forestry impacts. *Wetlands* 14: 1–9.

Roberts, B.H., R.J. Stimson, and R.R. Stough. 2006. *Regional economic development: Analysis and planning strategy*, 2nd ed. New York: Springer.

Rompré, G., Y. Boucher, L. Belanger, S. Côté, and W.D. Robinson. 2010. Conserving biodiversity in managed forest landscapes: The use of critical thresholds for habitat. *Forestry Chronicle* 86: 589–596.

Rushton, S.P., S.J. Ormerod, and G. Kerby. 2004. New paradigms for modelling species distributions? *Journal of Applied Ecology* 41: 193–200.

Sallenave, J. 1994. Giving traditional ecological knowledge its rightful place in environmental impact assessment. *Northern Perspectives* 22: 16–18.

Salmo Consutling. 2006. *Developing and implementing thresholds in the northwest territories: A discussion paper*. Yellowknife: Environment Canada.

Schaefer, J.A. 2003. Long-term range recession and the persistence of caribou in the Taiga. *Conservation Biology* 17: 1435–1439.

Scheffer, M., S. Carpenter, J.A. Foley, C. Folke, and B. Walker. 2001. Catastrophic shifts in ecosystems. *Nature* 413: 591–596.

Scheffer, M., J. Bascompte, W.A. Brock, V. Brovkin, S.R. Carpenter, V. Dakos, H. Held, E.H. van Nes, M. Rietkerk, and G. Sugihara. 2009. Early-warning signals for critical transitions. *Nature* 46: 53–59.

Schneider, R.R., J.B. Stelfox, S. Boutin, and S. Wasel. 2003. Managing the cumulative impacts of land uses in the Western Canadian Sedimentary Basin: A modelling approach. *Conservation Ecology* 7(1): 8.

Schröder, A., L. Persson, and A.M. De Roos. 2005. Direct experimental evidence for alternative stable states: A review. *Oikos* 110: 3–19.

Seip, D.R. 1992. Factors limiting woodland caribou populations and their interrelationships with wolves and moose in southeastern British Columbia. *Canadian Journal of Zoology* 70: 1494–1503.

Shultis, J., and R. Rutledge. 2003. The Muskwa–Kechika Management Area: A model for sustainable development of wilderness. *International Journal of Wilderness* 9: 12–17.

Simberloff, D. 1998. Flagships, umbrellas, and keystones: Is single-species management passé in the landscape era? *Biological Conservation* 83: 247–257.

Smith, R.J., R.D.J. Muir, M.J. Walpole, A. Balmford, and N. Leader-Williams. 2003. Governance and the loss of biodiversity. *Nature* 426: 67–70.

Squires, A.J., C.J. Westbrook, and M.G. Dubé. 2010. An approach for assessing cumulative effects in a model river, the Athabasca River Basin. *Integrated Environmental Assessment and Management* 6: 119–134.

Strimbu, B.M., and J.L. Innes. 2012. Framework for assessing the impact of human activities on the environment: The impact of forest harvesting and petroleum drilling on habitat of moose (*Alces alces*) and marten (*Martens* [sic] *americana*). *Biodiversity and Conservation* 21: 933–955.

Stuart-Smith, K., C. Bradshaw, S. Boutin, D. Hebert, and A. Rippin. 1997. Woodland caribou relative to landscape patterns in north-eastern Alberta. *Journal of Wildlife Management* 61: 622–633.

Swift, T.L., and S.J. Hannon. 2010. Critical thresholds associated with habitat loss: A review of the concepts, evidence, and applications. *Biological Review* 85: 35–53.

Thomas, C.D., A. Cameron, R.E. Green, M. Bakkenes, L.J. Beaumont, Y.C. Collingham, B.F.N. Erasmus, M.F. De Siqueira, A. Grainger, L. Hannah, L. Hughes, B. Huntley, A.S. Van Jaarsveld, G.F. Midgley, L. Miles, M.A. Ortega-Huerta, A.T. Peterson, O.L. Phillips, and S.E. Williams. 2004. Extinction risk from climate change. *Nature* 427: 145–148.

Timoney, K., and P. Lee. 2001. Environmental management in resource-rich Alberta, Canada: First world jurisdiction, third world analogue. *Journal of Environmental Management* 63: 387–405.

Usher, P.J. 2000. Traditional ecological knowledge in environmental assessment and management. *Arctic* 53: 183–193.

Walker, B., and J.A. Meyers. 2004. Thresholds in ecological and social–ecological systems: A developing database. Ecology and Society 9:3. http://www.ecologyandsociety.org/vol9/iss2/art3/

Walker, B., C.S. Holling, S. Carpenter, and A. Kinzig. 2004. Resilience, adaptability and transformability in social–ecological systems. *Ecology and Society* 9: 5.

Wallace, K. 2007. Classification of ecosystem services: Problems and solutions. *Biological Conservation* 139: 235–246.

Weckworth, B.V., M. Musiani, A.D. McDevitt, M. Hebblewhite, and S. Mariani. 2012. Reconstruction of caribou evolutionary history in Western North America and its implications for conservation. *Molecular Ecology* 21: 3610–3624.

Worm, B., E.B. Barbier, N. Beaumont, J.E. Duffy, C. Folke, B.S. Halpern, J.B.C. Jackson, et al. 2006. Impacts of biodiversity loss on ocean ecosystem services. *Nature* 314: 787–790.

# Chapter 4
# Cumulative Effects and Impacts: Introducing a Community Perspective

**Greg R. Halseth**

***With contributions by*** **David J. Connell, Dawn Hemingway, Derek O. Ingram, Glen Schmidt, and Karyn Sharp**

## 4.1 Introduction

As described in Chaps. 2 and 3, impact assessment has a long history and track record with respect to environmental issues. The assessment of natural resource development impacts on environmental systems is sophisticated, but it is also being challenged (Noble 2009). Critiques of contemporary assessment processes, as illustrated earlier in the book, have highlighted that cumulative impacts cannot be understood if our attention is limited to single stages of projects or is constrained to inappropriately small spatial or temporal scales (see Sect. 3.3 in Chap. 3). Further challenging assessment processes is the fact that social change is now more clearly and closely linked to the issues of environmental, community, and health impacts from natural resource development. Thus, a more integrated understanding of cumulative impacts cannot be obtained if it is limited only to environmental or natural resource topics.

This chapter builds on previous chapters by adding a consideration of community-related topics in the context of integrated cumulative impacts. It develops some of the critical community aspects needed to support a wider understanding of the cumulative impacts of natural resource development. The principles for understanding the complex and interwoven relationships among a host of large- and small-scale resource development activities over time and the pressures they create on communities speak to a much wider array of economic, policy, and community development undertakings. The purpose of this chapter is to add a consideration of community issues to the conversation around re-thinking an integrative regional cumulative impacts assessment framework.

G.R. Halseth (✉)
The University of Northern British Columbia, Prince George, BC, Canada
e-mail: Greg.Halseth@unbc.ca



M.P. Gillingham et al. (eds.), *The Integration Imperative*,
DOI 10.1007/978-3-319-22123-6_4

The chapter begins with background on the concepts of cumulative effects and impacts. It then explores the elements that would support a broader understanding of cumulative impacts that draws upon existing understandings of rural and small town community development. The chapter closes with a summary of the potential for incorporating community characteristics and considerations into a wider and more integrated understanding of regional cumulative impacts.

## 4.2 Background

This section of the chapter provides background information that will support an exploration of the potential to incorporate community issues into our understanding of the cumulative impacts from natural resource development. Such assessments should be conducted within the context of a robust framework, and a number of evaluative frameworks already exist for assessing project impacts. These include the components of existing environmental impact assessments (see Chaps. 2 and 3), social impact assessment (SIA) (Franks et al. 2011; see Boxes 4.1 and 4.4), and health impact assessment (see Chap. 5) processes. A recent study by Michell and McManus (2013) on how social issues have been and are being understood within the context of development of the Australian mining industry identified that SIAs

**Box 4.1 Social Impact Assessment: An Overview (David J. Connell)**

An SIA, like other types of impact assessment, is a systematic acquisition and evaluation of information in order to measure the positive and negative consequences of specific actions and interventions. These actions and interventions include a wide range of possibilities, such as industrial developments, public programmes, and government policies and projects.

Although different types of impact assessment are related to each other in some way or form, each one attempts to evaluate a situation through a particular lens. The social domain, which can be defined to include almost everything, can be operationalised in very simple terms; namely, SIA refers specifically to assessing impacts *as if people mattered*. Let's examine this further by considering how different types of impact assessment can be completed both without and with a consideration of people.

- *Economic*: Through an economic lens, one can look only at the number of jobs created and the amount of money circulated; through a social lens, one would also look at the quality of jobs on people's livelihoods, the impact on families if workers have to move, and the consequences of (in)equality of new versus existing wages.
- *Environment*: Through a biophysical lens, one can look only at things such as flora and fauna or hydrological cycles; through a social lens, one would

(continued)

**Box 4.1** (continued)

look at the impact on accessibility for recreation, visual quality objectives, and culture and traditional ways of life.

- *Health*: Through a health lens, one could look at factors such as the prevalence of disease, measures of body mass index, or longevity; through a social lens, one would look at quality of life, the impact on families, and social isolation.
- *Agricultural*: Through an agricultural lens, one could look only at soil capability and fragmentation of the land base; through a social lens, one would look at the quality of livelihood for a farmer.

Although it might seem odd not to consider the impacts on people, these examples speak to how social impacts are usually integrated seamlessly into other types of assessment. Such SIAs adhere to the same principles as other assessment tools and involve the measurement of direct and indirect impacts of both intended and unintended consequences. A valid and reliable measurement depends on specifying not only the outcomes, but also the intervention, as well as identifying the population, delineating the context, and operationalising what success/failure means for the particular action or intervention. Social impact assessments cover three scales of analysis: individual members of society, specific groups of people, and society as a unit. The latter can include geographic scales of neighbourhood, city, town, region, province, or country.

To anticipate and measure the potential positive and negative impacts of an action or intervention requires that SIA be oriented towards the future. Thus, an SIA is always limited by how certain we can be of what lies ahead. This uncertainty is compounded by the fact that studying the potential impacts of future actions on people is far more difficult than studying the impacts on the economy or the environment. Furthermore, to determine whether an impact is positive or negative one must have a standard or point of reference against which the impacts are assessed. It is essential, therefore, to include a clear conception of a desirable future society when completing SIAs. In this sense, SIAs are essential because they help guide society towards the common good of all people.

were only one component of a necessarily wider approach to community engagement and dialogue about the changes that specific places and larger regions are experiencing (see also Owen and Kemp 2013; Prno 2013). They also suggested that such a broader engagement and dialogue are increasingly crucial to the success of both community development and economic development endeavours (see also Markey et al. 2008). A recent report on the Australian experience with Liquid Natural Gas (LNG) suggested that:

> Building trust with the community is a key element in securing progress in gas developments. Differences in experiences in Queensland and New South Wales show that if widespread community trust is lost or never obtained, gas projects will not happen. (Grafton and Lambie 2014, p. 4)

Conducting SIAs may play a role in building trust and relationships, or these benefits may only be obtained after implementing and acting on the resulting plans. Will the social impacts identified in the SIA be accounted for by means of impact management plans? Many countries lack frameworks for mitigating the socioeconomic impacts associated with large-scale resource development projects (Shandro et al. 2011). Another major issue is that some project proponents and community stakeholders may feel that the process stops once the SIA is done. But in fact, this should be just the beginning of a long-term relationship. Structures and mechanisms must be developed to implement, monitor, adjust, and renew the plans that result from the SIA, and this includes the possibility of repeating the SIA in the new (possibly changed) context. Further complicating matters is the fact that SIAs, social impact management plans, and the associated impact mitigation strategies each come with varieties of different interpretations, meanings, and responsibilities assigned to them.

With respect to concern about understanding and evaluating the cumulative impacts of industrial resource development projects, four critiques have been made of existing environmental or SIA processes. The first is that these processes typically only examine or include topics and issues that have been identified through legislative frameworks, certification requirements, or current best practices (see Chap. 2). The potential for exploring new topics and issues within such assessments is not available given that they are generally time limited and purpose driven within specific project approval processes (see Chap. 3). Although such assessments can be valuable and may add important information, they do not contribute directly to understanding or evaluating the cumulative impacts of multiple projects over time; in addition, they often do not include a community perspective. As Palen et al. (2014) note, restricting frameworks for the evaluation of impacts and confining ourselves to incremental decisions about incremental projects forces us to function within a broken process that fails to consider larger and longer term implications.

A second critique, which derives directly from the previous critique, is that such assessments are typically bound to a single current project or proposal. As Noble (2010, p. 11) argued, "the problem is that many cumulative effects are not directly associated with the impacts of any individual project". As demonstrated in Chaps. 2, 3, and 5, the manner in which the boundaries are structured (physically and conceptually) around individual projects can have a significant bearing on the scope of the topics and issues that the existing environmental, social, and health impact assessment exercises are charged to explore. As Loxton et al. (2013, p. 52) noted, even single changes and interventions "interact and aggregate, and are influenced by additional interventions and exogenous factors, leading to cumulative social impacts".

The third, and perhaps most important, critique comes from the fact that existing social and health impact assessment exercises are not cumulative or historical in their design. Without attention to the impacts of other (past, simultaneous, and future) resource development activities, and without attention to how those various activities have transformed community characteristics and issues over time, existing exercises can only slightly increase our understanding of the trends and trajectories that are so important in community development work. Existing impact assessments rarely involve a broad range of stakeholders in their design. This can limit the scope of the topics, issues, and impacts that are tracked over time. Stakeholders such as industry, community organisations, and governments have also used different sources and types of information and methodologies to assess or describe the impacts of resource development projects. These design elements specifically limit the value of existing frameworks for assessing cumulative impacts.

A fourth critique is that most approaches to cumulative impact studies begin with an assumption that the impacts will be negative. This may be linked to the origins of such work in assessing the environmental disturbances caused by an industrial activity. But when considering community impacts, it is more difficult to assign normative labels such as *positive* or *negative*. It is the short-, medium-, and long-term consequences of these impacts that might allow observers to attach normative labels. In such assessments, it is necessary to link the expected changes to what the literature tells us might be the longer term impacts. For example, although the pressures on a local housing stock caused by a resource development project in the short term might yield negative consequences in terms of affordability and a possible spike in homelessness, in the longer term there may be a general improvement in the quality and variety of the local housing stock. Another example might be the disruption caused by new people moving to the community. In the short term, this may be negative, but over a longer term, it may become a benefit in the form of new ideas, new volunteers, and new skill sets becoming available and helping to build community resilience.

These critiques reflect a long and diverse suite of concerns within academic, policy, community, and business or industrial circles, as many individuals in these circles recognise that the current process is broken (see Chap. 7). Palen et al. (2014, p. 466) describe a need for "more transparent and comprehensive decision-making processes that incorporate trade-offs among conflicting objectives" that would in turn be assisted by new decision-support tools. Drawing upon one particular environmental conflict in BC, Cayo (2014) recently reported that Tom Gunton, an expert on environmental regulation, noted how "after a period of constructive dialogue between diametrically opposed interest groups, *people who'd barely speak to each other* worked out a solution that, although not perfect, was far better than any alternatives that seemed possible when the confrontation was at its peak". Inclusion and dialogue, as opposed to exclusion and judicial process, are important.

The need for decision-support tools and supportive, top-down public policy challenges the capacity and focus of most governments (Australian House of Representatives 2013). In Alberta, the provincial government formed a cabinet committee to respond to industry needs and the pressures that were facing rapidly growing communities (Government of Alberta 2006). The Oil Sands Ministerial Strategy Committee was supported by a Secretariat that included deputy ministers from across the government, who worked together to develop a common approach to assessing the infrastructure and socioeconomic impacts of resource development, to coordinate regional infrastructure plans, and to develop and coordinate community impact benefit agreements.

If our goal is to move towards a broader and more inclusive way to understand cumulative impacts, it is important to start with the recognition that there will always be unforeseen, unexpected, and unknown aspects. A broader understanding should be open to including such new aspects, particularly if they are discovered later in the process. Experience has taught us that many such aspects are discovered only over time and usually in a piecemeal manner (see Boxes 3.2 and 5.2). It is important to recognise the need to pull together those aspects that are known and which research and experience report are potentially likely. For these aspects, the cumulative impacts should not be a surprise. Instead, they should fall within the *scope of the possible* in any particular investigation.

Identifying, describing, and potentially predicting the range of multiple and interacting impacts from multiple industrial resource development activities that are all acting together through time and across space become even more complex as a result of the messiness of community processes and underlying values related to social, economic, and cultural issues. In this sense, a community-informed approach to cumulative impacts runs into the same problem of *causality* that is so challenging in other areas of research. To describe something as having an impact on some aspect of the community is not the same as proving causality in a court of law or in an application before an approval process. Kinnear et al. (2013) tried to move beyond simple causality by looking at both the direct and the indirect risks arising from a range of public health issues in the context of long-distance commuter labour in Australia's natural resource sector (see also Moran et al. 2013; Chap. 5). Our goal to broaden our understanding and dialogue is so that those who do approve resource development proposals should not be able to subsequently say that the impacts were unexpected even if the causes are shared. These issues make undertaking a conversation aimed towards broadening an understanding of the community aspects of integrative regional cumulative impact assessment both necessary and worthwhile.

## 4.3 Effects and Impacts

Noble (2010, p. 2) noted that cumulative effects and impacts are "one of the most perplexing issues in EA and natural resource management" and how "in the context of project based decision making, [the] current approach to CEA does not provide the results needed to understand broader environmental change or to make longer-term decisions concerning the sustainability of current and future development actions." As described in Chaps. 1 and 2, an effect is a change to the current circumstances caused by one or more resource development actions—something that may cause direct and observable changes. Individual effects may be more or less easily and clearly seen through the responses of the landscape or the community. For example, an upswing in oil and gas production requires additional workers, and that increase in the workforce (the effect) would be seen in the form of more people joining the community. The impacts of those additional people are likely to be wider than just a numerical increase, and will sometimes be more difficult to discern. In addition, such impacts are multiple, interwoven, complex, and changeable over time, and need to be traced at both local and (at least) regional scales. As noted in Chaps. 2 and 3, effects and impacts are complicated—there are direct and indirect effects and impacts, as well as cumulative effects and impacts, and there are also questions of thresholds and tipping points.

A number of research traditions and understandings exist for the community development side of resource industry projects, and they point towards a framework for interpreting those impacts. Pulling together those research traditions and understandings is the goal of this chapter.

Drawing on the literature, Noble (2010) identified a number of ways to understand the accumulation of individual impacts over time. These include linear additive, amplifying or exponential, discontinuous (see Box 4.2), and structural surprises. Although the first two are fairly easy to understand, discontinuous impacts are important because there may be no apparent impact until a certain threshold is reached, after which significant change can occur (see also Chap. 3; Kinnear et al. 2013). Similarly, structural surprises occur when there are multiple stressors or activities whose individual incremental levels of change do not appear to suggest the risk of a complete or new form of transition, but together, they result in a significant change.

**Box 4.2 Discontinuous Impacts (Greg R. Halseth)**

A simple illustration of discontinuous impacts can be found in the housing market pressures associated with resource industry booms or project construction periods. To start, the local housing market operates under its previous *status quo* conditions. When a sudden upswing in activity occurs, new industry or construction workers move into the community to support that increase. Early in this process, most of these workers will be absorbed into the existing vacant rental or homeownership housing stock. There will at first appear to be limited impacts within the housing market. However, once the vacant housing stock is consumed, rents and prices will tend to increase rapidly due to competitive market pressures. Some landlords will undertake evictions in order to bring in new tenants who will pay a higher rent. As a result of these actions, long-time residents living on limited incomes will face higher costs and may be squeezed out of the housing market (Ryser and Halseth 2011). Homelessness and poverty can suddenly appear even though the community is experiencing what many would call positive economic development. The workload on volunteer groups will increase to deal with this spike in local poverty. For example, in addition to a local food bank, other food sharing and meals programmes may have to be established to meet the growing need for their services. As these pressures and demands rise, volunteer burnout may increase and some of these key community social services may collapse. Therefore, impact assessment is not just about economic planning or environmental remediation; it is also about the critical need to include social planning and community development investments.

Noble (2010) also talks about the complexity of pathways and the synergistic nature of cumulative environmental impacts. As the authors of previous chapters noted, cumulative impacts are generally obvious when looking back in time, but are more difficult to assess looking forward. This is made more challenging at the local and regional level. For example, due to short-term and limited funding options, important local organisations may close, and the resulting loss of institutional memory about historical issues can make it difficult to shape our understanding of cumulative impacts.

## 4.4 Exploring the Community Aspects of Cumulative Impacts

This section focuses on describing and outlining a structure for reviewing the potential of cumulative impacts from a community perspective. It draws upon most of the constituent elements described in the literature that help to structure and organise communities. Although current assessment processes use aspects of the elements described herein, it is critical to remember that for the evaluation to be integrative, regional, and cumulative, it must include attention to the multiple resource development activities that are occurring and that are interacting over time and across space.

In their review of environmental impact assessment approaches, Peterson et al. (1987) identified a series of circumstances that are important for understanding and managing cumulative impacts. They highlighted the need for cooperation across jurisdictional boundaries and the need to be inclusive of a host of interested parties outside of government regulatory bodies. Furthermore, they highlighted some persistent challenges with current approaches. These included the problem that indirect impacts, which take longer to manifest, and impacts that arise outside of well-understood *pathways* are more challenging to understand. These circumstances, needs, and persistent challenges nicely presage the issues inherent in trying to bring a community perspective to the understanding of cumulative impacts from natural resource development.

Nitschke (2008, p. 1715) emphasised that the "rapid acceleration of industrial development in north-eastern BC is currently occurring without a comprehensive assessment of the effects it will have on ecological or cultural systems". With a specific focus on Canada's Treaty 8 territory (parts of northeastern BC, northern Alberta, northwestern Saskatchewan, and southern NWT), he argued that the connections between environmental, community, and health issues that are broadly understood from within an Aboriginal cultural framework (see Chap. 1) have not been incorporated into cumulative impact assessments.

In a review focused specifically upon First Nations involvement in EIA processes across BC, Plate et al. (2009) found significant frustration among these communities. These frustrations arose from how the terms of reference were created, how the timelines and processes did not support Aboriginal decision-making processes, and how the participants in some processes appeared unwilling to consider many of the values that Aboriginal people consider important. Additional concerns included a lack of process transparency and consistency, as well as the actual mechanics and costs of participation. They also highlighted the way that "some Project proponents who were unenlightened about First Nation rights and interests, or who merely see First Nation participation as another obstacle to overcome in the pursuit of their Project" can limit the opportunity for effective dialogue and building of relationships (Plate et al. 2009, p. i). Their report goes on to highlight 75 best practices that could be better incorporated into contemporary assessment processes so as to support the involvement and viewpoints of First Nation communities.

**Box 4.3 Ground Zero: Impacts of Natural Resources Development on First Nations Communities in Canada's North (Karyn Sharp and Derek O. Ingram)**

The current boom in oil and gas exploration, and specifically LNG projects, is having a significant impact on First Nations communities across Alberta and BC. Although many projects are still in their early stages, First Nations are being greatly affected by the exploratory infrastructure development of gas and oil deposits upstream at the source, by midstream transportation, and downstream at the coastal terminal. Despite the lack of direct gas development in central BC to date, the pressures of project development are having significant impacts within the communities. Some of these, including rising tensions and conflict, are fed by feelings of anxiety, anger, frustration, confusion, and optimism, and are exacerbated by low capacity levels within Nations.

First Nations communities in Canada are directly affected by much of the national economic development and growth, and have factored into, willingly or not, most of the national economic growth strategies since the fur trade of the 1600s. However, economic benefits to First Nations have historically been arguable and the current push for oil and gas resources expansion is no different.

Currently, there is strong provincial and federal support for increased expansion of mineral, oil, and gas resources, as there is a general national consensus that natural resources are, at present, Canada's best economic potential. The pace of natural resource extraction in Alberta and BC is on the rise again, the likes of which have not been encountered since the 1950s and 1960s. The intensity of resource extraction (e.g. forestry, mining, and hydroelectric developments) is expanding. New technologies and expanding infrastructure development are encroaching on new areas previously untouched; minerals are being extracted from further below the Earth's surface than ever before, while forestry is expanding the harvest base due to the devastation caused by the mountain pine beetle. The economic advantage to a company to be the *first one out of the gate* on a project results in rapid-paced project planning and intense pressures on First Nations communities.

From 2013 to 2015, the First Nations in central BC have had to deal with five proposed LNG projects, some with multiple pipeline projects in their

(continued)

**Box 4.3** (continued)

territory. These projects each requires engagement with the different companies proposing the projects, requiring hundreds of meetings a year with industry representatives, internally with the Nation members and staff, and between the Nations all just to keep informed about the proposed projects. In addition, traditional land-use studies, interviews, mapping projects, report writing, and review of technical data are required for each project. All of this work requires a small, well-staffed and well-funded army of individuals with the appropriate technical skills necessary to make informed decisions. Unfortunately, this level of capacity is almost impossible to muster in any given Nation for a single project, let alone four additional projects. This also does not take into account the hundreds of other proposals (forestry, mining, road, construction, fish, wildlife, etc.) that each Nation must deal with annually, which have inadequate, or no, natural resource referral systems and are already understaffed.

The First Nations communities whose traditional territories have oil and gas reserves, pipeline routes, timber-harvesting areas, and mineral deposits have to try to find balance within their communities between incoming opportunities and wealth, and their traditional cultural values. For First Nations leaders, decisions are difficult in part because the low levels of capacity do not allow for adequate understanding of, or engagement in, the proposed large-scale LNG projects. These pressures also do not provide adequate time for the Nations to determine how to reconcile their traditional needs with those of the modern economic reality. The desire for economic prosperity and stability by a Nation, through opportunities like LNG development, often comes with unanticipated consequences. These consequences can include increased social diseases like alcohol and drug abuse, the abandonment and/or dilution of traditional values, and the loss of the lands themselves.

It is critical that First Nations communities have the time and the capacity to make thorough decisions that are value laden for the future health and welfare of their respective communities. First Nations communities need: adequate time to examine their future needs and goals that often include physical, mental, and spiritual health; education programmes for youth and adults; and infrastructure that meets the Canadian norm to name just a few. However, current industrial opportunities that flow through these communities can also be the impetus for addressing gaps, like community health care and those named above, with mechanisms like revenue-sharing opportunities, and economic opportunities through business development, employment, and education.

Significantly, the disparity between Canada's communities needs addressing. The dichotomy of capacity levels for a Canadian community versus a First Nation community in Canada is exceedingly broad. Stark inequalities in capacity hamper First Nations communities and organisations across Canada. Vast majorities of industry-affected First Nations are, therefore, forced to recruit and hire outside (often non-native) experts and consultants in order to properly assess and manage development requests. Further perpetuating the cycle of lessened capacity within

(continued)

**Box 4.3** (continued)

First Nations communities is the dilution of internal community energies, specific to economic development opportunities, that come from budgetary cuts and more pressing matters such as housing; there is little time to focus on capacity building when dealing with the hurried time lines put forth by project proponents.

First Nations people are not against most development projects. They recognise the economic benefit to the local and global economy, but their participation in that system is at best superficial and at worst unsupported. The question then becomes how then do First Nations address and manage these project proposals and ensure their involvement while considering the goals of their Nation in the long run?

Rapid economic growth and the utilisation of natural resources on First Nations traditional territories can have serious long-term repercussions. Specific to the oil and gas industry, the changes, often negative, to First Nations communities range from social, cultural, and health to spiritual. These impacts (risks and benefits) need to be examined thoroughly before land-altering decisions are made by Canadian governments on First Nations territories.

Exploring cumulative impacts means that it is necessary to understand a range of contexts. These background contexts include the scope and extent of social, economic, demographic, political, and environmental change. For First Nations communities, cultural, spiritual, and heritage considerations are also important (see Box 4.3). In addition, project proponents often impose a rapid pace for introducing new economic activities into a region that draws primarily upon natural environment resources. Third, there is the accumulated legacy of past economic activities that also drew upon that natural resource base.

The notion of a *legacy* deserves some additional attention (see Box 4.5). At one level, past development is likely to have produced a host of impacts that have been acting over time and across the region. The degree to which the historical confluence of these impacts may be recognised by contemporary communities or assessment processes, and the degree to which an adaptation trajectory may be entrenched as a result of a confluence of historical impacts, is important in any debate about development options and future visions for a community, region, or economy. How these legacies collectively interact with the broader aforementioned economic, social, demographic, and political changes in communities and regions is a critical matter of concern in order to understand the potential magnifying and multiplying implications of proposed new developments.

There are also some inherent tensions in these contexts as a result of our desire to understand cumulative impacts. For example, sustainable revenue flows are affected by, and potentially reduced by, impacts. There is a tension between the immediate requirement of households, communities, and economies for jobs and income to support daily needs and the long-term need for a more sustainable community and a more resilient economy. There are also the tensions inherent in determining who obtains the bulk of the benefits from a natural resource-based economic activity (see also Box 4.6) and who bears the costs. Too often in rural resource development, it is those outside the local community or region who gain the benefits while the community and region bear both the immediate and the long-term costs of the activities (Government of Alberta 2006; Morris 2012). Exploring how the inclusion of a wider consideration of environmental, community, and health issues can contribute to a broader understanding of emergent cumulative impacts may end up highlighting many more important inherent contradictions and challenges.

### 4.4.1 Assets and Aspirations

Economic transactions have become faster and the economy has become more connected than ever before (Harvey 1990, 2005; Ryser et al. 2014). These changes have also been shaped by increasingly rapid fluctuations in commodity prices. Booms develop faster; busts become deeper. In many communities, the process of understanding, planning for, and responding to the cumulative impacts of resource development has become more complicated as community stakeholders, service providers, and community groups must increasingly respond to growth in one resource sector and a decline in others. Community development, however, is about building the capacity of people and groups to marshal information about both short- and long-term change so they can respond proactively and meet both challenges and opportunities on the community's own terms (Kretzmann and McKnight 1993; Markey et al. 2008; Lo and Halseth 2009; Reimer 2010).

The argument has been made that within the global economy, technology is decreasing the relevance of distance. As *space* has become less of an issue in economic development, *place* is becoming more important (Markey et al. 2012). To respond to the myriad community, social, economic, cultural, and environmental changes that are occurring, communities need to realistically consider and enumerate their *place-based* assets. Then, they need to creatively re-imagine these assets and re-bundle them to create new competitive advantages for their place and region. How they do this must be linked to their aspirations as a community vis-à-vis the type of place, society, and economy that they wish to create (Halseth et al. 2010). Large natural resource development projects, especially those guided by companies from outside the community or by the needs of government policy, have the potential to disrupt local considerations of both assets and aspirations.

### 4.4.2 The Challenge of Causality

The long and complicated public and legal debates that arise around how some products or activities may affect human health is an important reminder of the challenge of causality that looms over all cumulative impact work. Many times, these debates are framed around a human need to find and allocate blame. Sometimes, they are also linked to legal proceedings for claiming financial and other forms of compensation for damages.

Causality is often clear, as in the cases of the direct discharge of untreated industrial effluent into bodies of water or deforestation of steep slopes with no effort to prevent erosion of the soil. The impacts of such activities are clear and easily accounted for. More difficult is the challenge of dealing with subtler impacts. In community development work, and with our interest in assessing the cumulative impacts of multiple large and small resource development projects over time, the focus is not on proving causality (see also Chaps. 5 and 7). If there is anything the debates noted above have taught us, it is that there are many intervening factors that complicate the establishment of causal relationships. Rather than focusing on causal pathways (i.e. linkages), the need is to instead focus upon the reasonable expectations about individual, community, and regional impacts to help prepare for, and (where needed) mitigate, various impacts, as well as how to identify ways to take advantage of the potential new opportunities that may be presented.

The focus on elucidating reasonable expectations around cumulative impacts across the spectrum of community development issues should also be linked to differences in the community's coping capacities. The ability to cope is, of course, partly related to the magnitude and duration of the activity and its impacts. It is also related to the community's previous experiences with exercising their coping skills and resilience. As well, community systems may be more or less resilient as a result of past community development and change processes that have either damaged or improved their capacity. The elements of community development described in Sect. 4.7 in the context of social capital and social cohesion are especially important in determining a community's resilience and ability to cope with various impacts.

## 4.5 The Scope of Issues to Consider

Our focus is on how cumulative impacts from multiple natural resource projects change communities. As Krzyzanowski and Almuedo (2010, p. 1) noted, cumulative impacts describe "spatially or temporally accumulated changes that result from the perturbations of one or more resource sector activities". To begin an exploration of the scope of the issues that need to be considered, this section explores two elements of scope: the geographic scale and the temporal scale.

### *4.5.1 Geographic Scale*

Cumulative impacts can vary as a function of the geographic scale at which they are examined (see also Chaps. 5 and 6). From a community perspective it is useful to highlight four *geographies*: the individual, the household, the local community, and the regional community.

#### 4.5.1.1 Individual

The individual level is the most intimate geographic scale at which to examine cumulative impacts (Bakke et al. 2004; Van Hinte et al. 2007). Following changes in employment (for example), people feel the stress of that change, the hope of opportunity, the challenge of loss, and other emotions. The range of impacts can be emotional, financial, psychological, and physical. Emotional impacts can be a response to change, to fear and worry, and to conflict related to that change. Financial impacts reflect one's capacity to meet the instrumental needs of daily living. Concern about one's financial capacity to pay for shelter and food, and to maintain a reasonable quality of life, can create significant stress for individuals and families. Psychological impacts represent how people respond when their aspirations are not realised, or when the goals that are part of their quality of life are challenged (see Box 5.4).

By way of example, consider the individual who values access to natural wilderness areas for recuperation and *re-creation*. This value is one that holds many people in rural and small town places. Individual projects in natural areas may not only cause short- or longer-term impacts, but may often lead to small-scale closures of public access to wilderness areas that are important to such individuals. When multiple projects are executed over a wider territory and a longer time span, however, the impacts are greater because this may effectively close off access to large tracts of wilderness. This significantly interferes with one of the key reasons why people live in that area. The results can have important impacts on the physical and emotional well-being of individuals (see Vignette 6.5). The magnitude of these impacts depends on the resilience of individuals (i.e. their ability to cope with emotional, financial, psychological, and physical manifestations).

**Box 4.4 The Social Impacts of Oil and Natural Gas Development in Northern British Columbia (Dawn Hemingway and Glen Schmidt)**

An economic boom due to rapid development and expansion of the hydrocarbon industry does not occur without social consequences. Economic growth is generally accompanied by population growth in the affected communities. This exerts pressure on the housing stock because the supply may not meet the increased demand while newcomers compete to find a place to live. A situation like this creates particular challenges for existing residents who may be economically marginalised due to factors such as poverty, disability, and age. For example, people with severe and persistent mental illness and people with intellectual disabilities may find themselves pushed right out of the housing market. The current maximum shelter rate for a person with a disability in BC is $375.00 per month (Disability Alliance BC 2014). In a market where housing prices and rental costs are rapidly rising due to limited supply it is easy to understand that a person with a disability will struggle to find shelter. The

(continued)

**Box 4.4** (continued)

same is true for seniors who may live on a small fixed income. A rapid rise in the cost of living, especially housing, can render a senior destitute.

A sudden increase in population also creates a gap in services for other population groups. Young families that move into a rapidly expanding community may face shortages in services like licensed daycare facilities, schools, and recreational outlets for children. Newcomers have few pre-existing supports to turn to for assistance such as family and close friends. The absence or shortage of public services creates a precarious situation, as a young family that faces any form of crisis may have nowhere to turn. Close friends and family may be far away and the social service infrastructure may not have yet caught up to the population growth. Stress and worry about the environmental impact of oil and gas enterprises is another challenge that may affect physical and mental health.

People who work in the oil and gas industry tend to have high incomes. The experience in communities such as Fort McMurray, Alberta, or Fort St. John, BC, suggests that high disposable income and a relatively youthful population are also associated with increased problems of substance abuse and misuse. Services and treatment options may not be adequate to meet the demand and this results in a variety of situations. Incidents of child abuse, woman abuse, motor vehicle accidents, and legal issues may increase in number and frequency.

Although job growth and job opportunities are good things, in the context of a boom there are also negative consequences. Limited services, housing shortages, substance misuse, and increased marginalization of people who are already marginalised represent a few of the factors that need to be considered when looking at rapid economic expansion. Measures are needed to not only ensure that the development process is environmentally sound and sustainable, but also that a portion of the wealth created is allocated for social infrastructure within the regions where exploration and extraction occur.

#### 4.5.1.2 Household

Moving up the spatial scale of impacts, households can experience a similar set of impacts (Van Hinte et al. 2007; Perry 2013). Without going into the same detail, these are connected to issues like job loss, changes in job responsibilities, and changes to personal resources, all of which can produce impacts such as a fear of change, conflict over change, financial impacts, and decreased quality of life. Moreover, these impacts can affect the strength and quality of the relationships among individuals in a household, creating further impacts. In addition, both individuals and households are affected by changes in local public and private sector services. A few job losses at a single plant or mill may not necessarily affect local public services. But when large numbers of jobs are lost through rationalisation and automation, and are combined with government policy efforts to reduce debt and deficit by closing or regionalising public services, the combination can have a significant impact on the local services that are available to support and renew the capacity of individuals, households, and communities (Halseth and Ryser 2006; Sullivan et al. 2015). In time, the loss of public and private services can affect the ability of some households to live in the community. Thus, when evaluating cumulative impacts, it is necessary to know the types of demands for and the expectations of local services and support by current and future populations so they can respond more effectively to changes resulting from resource development.

#### 4.5.1.3 Local Community

The next larger geographic scale related to the cumulative impacts of multiple natural resource development projects is the local community. For many, this is the only scale that comes to mind when thinking about cumulative community impacts (Fidler and Noble 2012; Gachechiladze-Bozhesku and Fischer 2012; Arce-Gomez et al. 2015; Wong and Ho 2015). At one level, community is affected by multiple instances of the individual- and household-level issues noted above, which accumulate as the number of affected households increases. As noted in Sect. 4.4.2 (The Challenge of Causality), a local community's capacity to cope with cumulative impacts, and to be resilient in the face of those impacts, depends upon all the issues noted in this section and by the community's past experience with the need to adjust and cope.

An important point to add at the local community scale concerns the issues of *incrementalism* (the accumulation of impacts) and *tipping points* (thresholds beyond which sudden changes occur to a new state). In considering the cumulative impacts of multiple LNG construction projects, for example, are both phenomena at work. If one plant is under construction, there would be a noticeable impact on local healthcare services because many of the construction workers will not be living in organised camps, but rather in the community. As a result, they will not have access to camp-based primary health and wellness support. Although a local health centre and its staff may be able to cope with some incremental increase in the demand for their resources—something that is not guaranteed—the number of new users of the local health centre will increase as other projects start until the number reaches a tipping point, leading to crisis when the health centre can no longer support its community. This topic is related to the discontinuous impacts described by Noble (2010). Again, it is necessary to know the types of demand for local services and support throughout the period of different projects and how such demands may become cumulative.

#### 4.5.1.4 Regional Community

The largest of the geographic scales is the regional community (Lockie et al. 2009; Moran et al. 2013). As Krzyzanowski and Almuedo (2010, p. 43) note,

> [the] more sectors or resource activities that can be included in a cumulative impacts assessment, the better, and, in theory, the more realistic its results. Increasing complexity, however, may amplify the error associated with these evaluations. Regional assessments are suggested because impacts are far-reaching.

Northern BC, for example, is organised as a set of nested communities, subregions, and regions, and these nested units function together. Cumulative impacts within a single community have carry-on impacts on neighbouring communities within each regional system. For example, the development of large retail shopping complexes on the urban edge of a regional centre will create a range of local impacts as the retail sector adjusts to both opportunities and challenges. It will also create sets of impacts on the retail sectors of adjacent communities, where retail leakage may affect the viability and availability of local services.

A second issue concerns the location of the resource development activities. If the development sites are outside of the municipal boundaries, town governments may have limited ability to tax them to obtain new revenues that will let the town respond to the impacts of those projects (e.g. to hire more support staff). The Fair Share Agreement developed for the Peace River region of northeastern BC is one response to such boundary and jurisdiction challenges: it creates a regional revenue-sharing arrangement based on the expected impacts of development (Markey and Heisler 2011).

In any attempt to understand cumulative impacts, it is important to consider all four geographic scales in the assessment. This attention to different regional scales is considered in the literature on *new regionalism*, in which the interplay between local and regional impacts is central to understanding the complexity of development processes (Massey 1995; Savoie 1997; Porter 2000; Hudson 2005). The origins of new regionalism are linked with the broader economic transformations of globalisation, an increase in regional disparities, and a breakdown of policy levers around investment and equalisation (Polèse 1999; Scott 2004). However, this approach builds on the notion that the region is a manageable scale for understanding impacts and designing mitigation strategies. Attention also extends to the discussion of *governance*. In Sect. 1.7, governance was described as the processes whereby societies or organisations make their important decisions, determine whom they should involve in those processes, and decide how they render account. Issues arise as localities and regions experiment with different institutional structures and relationships of governance to cope with their changing situation (Storper 1999; MacLeod 2001).

### 4.5.2 Temporal Scale

Scale also relates to the time frames around which cumulative impacts accrue and how long it takes them to work their way through the various components of a system. Three issues are particularly important in this context. The first has to do with how far back to go to establish a baseline for assessing cumulative impacts. In some cases, such as for BC's *instant towns* (typically, towns that were built at the same time as, and to help support development of, new resource projects and which have full municipal status), there is an almost exact starting point for the community that provides a logical basis for starting to evaluate impacts (Halseth and Sullivan 2002). In other cases, there may be a significant event or moment in time that helps to define a reasonable place for setting a baseline. But in many cases, it is difficult to choose an appropriate starting point because communities are always changing and adapting.

Included in the temporal scale is the time necessary for including both the *shock* to a system and that system's recovery. This is commonly described in the boom town literature associated with 1970s' oil and gas development in the USA and elsewhere (Freudenburg 1981, 1984; Gartrell and Krahn 1983; England and Albrecht 1984; Malamud 1984; Bone 1992; Brown et al. 2005). In these cases, the sudden increase in activity creates a shock to the local community system that includes all of the types of impacts described in Sect. 4.5.1 for the different geographic scales. As the development processes move from the intensive exploration and construction stages into lower intensity production stages, the local community system attempts to re-balance at a new equilibrium of activity and expectations.

A more general challenge for resource-dependent communities responding to shocks, booms, or busts is the degree to which these communities are *embedded* in the social, political, economic, and institutional structures of an older resource-based economy. Debate and theory on natural resource development, from staples theory (Watkins 1982; Drache 1991; Haley 2011), in which development is explained in terms of the extraction and export of minimally processed raw materials to more advanced industrial economies, to evolutionary economic geography

**Box 4.5 Legacy Impacts (Greg R. Halseth)**

It is a challenge to agree on just how far back in time one should go to establish a baseline for assessing cumulative impacts. Consider, for example, Aboriginal communities in northern BC. There is no doubt that they have been affected by multiple forms and types of natural resource development projects organised by non-Aboriginal interests for well over 100 years. Industrial resource extraction has accelerated these impacts during the past 50 years. It is possible to trace the start of particular industrial forestry, oil and gas, or mining activities back to specific policy or investment dates. But how do older processes such as the establishment of residential schools (see Vignette 6.6) affect the past and present individual and community capacity to cope with complex evolving situations and resilience to deal with the associated impacts? Indeed, even older processes such as colonialism and deterritorialisation are continuing to have impacts. It would be unfair to evaluate the contemporary pressures of change and their resulting multifaceted impacts without reference to those historical contexts (see Box 5.4).

(Martin 2000; Boschma and Frenken 2006; Boschma and Martin 2010), which focuses on the processes of change and innovation to the embedded routines, policies, and practices of particular modes of production, has highlighted the limits imposed by historic *path dependency* on places, regions, and states. For Innis (1933), staples theory described the social, political, and economic implications of Canadian resource development at national, provincial, regional, and local scales (see also Barnes 1996). The resultant social–technological–institutional systems that develop to support the export of raw resources (or staples) also develop social, political, economic, and institutional rigidities that resist change. On the other hand, a central debate in evolutionary economic geography is over the degree to which the embedded routines of such social–technological–institutional systems may be able to break from their past path dependency to innovate, change, and transform, or even to fade away and be replaced by a new economy (Page 2006; Martin 2010; Drahokoupil 2012; Oosterlynck 2012).

A second aspect of temporal scale relates to economic and community development issues. If there is to be a change in the level of natural gas exploration around a community, for example, how much and to what degree should earlier forms of gas exploration (and other natural resource projects) be included in an assessment? Is there a natural cut-off point where previous projects should or should not be included in an assessment of contemporary and future outcomes of cumulative impacts? One of the axioms of community conflict research is that local conflict and debate is rarely about the issue at hand, as it also tends to be influenced by past conflicts, debates, and competitions between leaders, egos, organisations, and philosophies (Halseth 1998; Lockie et al. 2009; Wong and Ho 2015). Another important question is how long legacy investments should be tracked or considered in terms of their impact. Should legacy investments (e.g. infrastructure) be tracked until they are replaced, regardless of their age and regardless of the changing community context?

This, of course, brings us to the issue of the future, and how far one should project the impacts of resource development in order to understand how this process will work out over an appropriate lifespan. A simple example is the coming of large forest-product mills to the small towns of northern BC in the 1970s. This occurred in the context of the province's industrial resource development policies and strategy (Markey et al. 2012). Early in that period, cumulative impact concerns at the community level focused on the needs of young families and children. Over time, the relatively stable work that these mills provided meant that young workers began to age in place (Hanlon and Halseth 2005; Hanlon et al. 2014). Now, 30–40 years later, the impacts of this policy and the resulting economic investment means that the communities are being forced to deal with the needs of an older population, including changes to health care, housing changes, and school closures. In another 10–20 years, the community will need to not only grapple with a greater demand for higher order services for seniors, which will continue to grow as the baby boom population ages, and with recruitment and retention issues to establish the next generation of the workforce—which requires attention, once again, to the needs of young families with children, but who will have a different set of expectations of the community and the economy than the families who were recruited in the 1970s. Further complicating matters is the rise of long-distance commuting by workers and the impact of *mobile lifestyles* on local worker recruitment and retention (Australian House of Representatives 2013; McCreary 2013). One impact is that a mill that opened in the 1970s created population, demographic, service, and infrastructure impacts that evolved over the following 50+ years.

In this context, the cumulative assessment framework has been shaped by industry actions (i.e. the announcement of a major project) and a response by those who will be affected. It can be argued, however, that the impacts start with a community's preparation to develop the next-generation workforce who will engage in resource development projects. The preparation, assets, and capacity that are in place will shape how the community is positioned to respond to the challenges and opportunities from resource development and, thus, will shape the impacts of that activity. Like ecosystems, communities will continue to transform in response to impacts, so continuing attention to planning for these changes, as they work through a system over time, will be important.

## 4.6 Issues Related to Community Infrastructure

In community development work, it has become clear that it is necessary to invest in four key sets of infrastructure (Markey et al. 2012; see also the multiple-capitals framework of Emery and Flora 2006): the physical infrastructure, human infrastructure, community infrastructure, and economic infrastructure. As the subsequent sections will show, the *directionality* of these investments will depend on the legacy of past investments, the scope and scale of the cumulative impacts, the coincidence of need and opportunity, and the community's own understanding of its assets and how they fit with their aspirations for the kind of community and economy they desire for the future. For each of these infrastructure topics, there will be pressures, and yet these pressures may lead to new opportunities. For example, as noted earlier, an influx of new people may strain the resources of a small community's volunteer sector, but over time these new residents may become a new source of volunteers or community leaders, or they may bring new ideas and innovations to help the volunteer sector take the steps needed to continue meeting their mandate.

### *4.6.1 Physical Infrastructure*

The physical infrastructure has two main components (Nielsen and Elle 2000; Harriman and Noble 2008). The first refers to those elements necessary for the *old* economy, and includes roads, railways, and basic physical infrastructure such as water and electricity services. The cumulative impacts of resource development projects can cause a great deal of wear on the existing physical infrastructure due to increased demand. A significant challenge is that the revenues to renew and redevelop that infrastructure may only come late in the development process, whereas the need to invest and reinvest may arise early in that process. Lessons about infrastructure investing from the 1950s to 1970s industrial resource development era in BC suggest that although roads may have been costly to construct, that infrastructure foundation has supported multiple economic booms in mining, forestry, oil, and gas for more than 50 years. It seems that the initial spending has produced a very good return on investment.

However, physical infrastructure also includes things that are necessary in the *new* economy. In particular, this means the infrastructure needed to rapidly move information, goods, and people. For many, the hallmark of the new economy's physical infrastructure is high-speed Internet service. Again, opportunities to use resource development construction projects to pay for long-term physical infrastructure improvements are a possibility. Increasing use of airports during the construction phase of projects can provide the fees, for example, to upgrade that facility so that it is better able to serve the needs of the community and region when airport use settles back to the operations level for the projects.

Cumulative impacts on physical infrastructure may be incremental, but they can also have tipping points. For example, increased oil and gas exploration can greatly increase road use and congestion. At some point, this can change from a minor inconvenience to a serious barrier or hazard for both residents and tourist traffic. As already noted, the fiscal resources to pay for needed improvements may not be forthcoming at the time when the improvements are actually needed. Although pressures to upgrade physical infrastructure may be driven by upswings during the construction booms that accompany resource development projects, it is important that (if they are built) they be planned at a regional scale and within the scope of expectations for the long-term demographic and economic structures of communities and regions during the much longer operations phase that begins after the new resource industries have begun their operations.

### 4.6.2 Human Infrastructure

The evaluation of cumulative impacts is perhaps most complicated for the *human infrastructure*. To start, a number of local organisations may find it difficult to participate in assessment processes because they lack the means. Tracking cumulative impacts is particularly difficult for the many volunteer and non-profit sector organisations that lack information management systems, including the resources to manage those systems (Ryser and Halseth 2014). Where information management systems do exist, issues concerning confidentiality and competition between organisations can impede the synergy and collaboration that come from sharing data about needs, demands, and trends.

As noted in Sect. 4.5.1, individuals and households can be affected in a range of emotional, financial, psychological, and physical ways. Fortunately, there is an opportunity to build infrastructure capacity through natural resource development projects. There are pressures to increase local employment early in such projects and these can provide the opportunity to develop leadership and problem-solving skills, as well as a host of other valuable and transferable workplace skills. Building that skills capacity, however, can be challenged by the movement from vertical to horizontal organisational structures in many industrial work environments; these have, in some cases, provided fewer opportunities for incremental growth and development. Looking further at cumulative impacts, it is important to consider the unintended consequences of changing government policies that have altered educational, employment, and apprenticeship programs. At times, there can also be a mismatch between gaps and shortages in the workforce and the focus of government programmes that provide training or retraining (e.g. for older workers affected by industry closures). Some government programmes only provide short-term funding, or only fund specific types of training, that will not allow residents to fully develop the job skills that are now demanded by industry. All of these issues, when combined, can have a considerable impact on the human infrastructure.

### 4.6.3 Community Infrastructure

When the community development literature speaks about investing in community infrastructure, it generally refers to the range of volunteer organisations and community-based groups that are active locally (Esteves and Vanclay 2009; Ryser and Halseth 2014). Such groups play an increasingly important role in delivering services. This role is accelerating as a result of reductions in the services delivered by private-sector or public agencies in many rural and small town places. Although they may deliver social, recreational, and cultural services and facilities, it is in the areas of health and social services that the cumulative impacts are most intensely felt. The challenge is that volunteer and community-based groups in small communities have (by definition) a small base of local people from whom to draw for their participants and volunteers. The failure of government services to expand to meet the increased demands when new resource development projects begin often puts additional pressures on volunteer groups to fill the gap. As the cumulative impacts of multiple projects begin to accumulate, there is a risk of burnout within these groups.

### 4.6.4 Economic Infrastructure

Consideration of economic infrastructure investments has at least three levels that need to be explored (Ryser and Halseth 2010): the enterprise, the local economy, and the economic coordinating and planning organisations. At the enterprise level, opportunity is always coupled with risk. Not all enterprises have the knowledge required to get involved in natural resource development. This is particularly the case if a new sector is coming to a community and no enterprise has experience with that sector. Cumulative impacts from mixed messages (e.g. whether or not the project will go ahead; whether none, one, or many projects will go ahead) make enterprise-level decision-making even more of a challenge.

The pace and scale of opportunities can sometimes overwhelm the local economic structure and put considerable pressure on the system to expand or grow. As more projects come online, cumulative impacts can increase stress on the entire local economic system. Individual economic sectors, such as the supply and support sectors, will feel particular pressure. Similarly, attention must be paid to the scope and breadth of the local economy, as well as the age of business owners and the need for attention to business succession planning and renewal. Local and non-local enterprises, as well as franchises, all act to complicate the structure of the local economy. The key to success is to build a capacity to capture wealth locally. The cumulative impacts of a large number of projects operating in a relative short time

**Box 4.6 New Jobs Can Limit Local Business Benefits (Greg R. Halseth)**

One example of how cumulative impacts can extend through the local economic infrastructure has to do with human capital—that is, the competition for workers and staff, especially in cases where there are shortages of either. One example involves a small community working with a major natural resource project. Early in the dialogue with the resource development company, a hire local plan was proposed whereby the company would try to reduce local unemployment levels as it brought new jobs into the community. The resource project needed some key individuals to assist with its operations. In the end, the project ended up hiring the local bus driver, the local government's accountant, and the local economic development officer, among others. The key trait of these individuals was that they were the only ones in the community with the specific training or certification required to do these jobs. The problem for the community was that there were no other local people who could drive the school bus, manage the local government's finances, or work to attract jobs to the community for the still large numbers of unemployed people who did not have these special qualifications. Even though this individual project had laudable aims, it significantly disrupted local capacity as those who had been working within a community structure now found themselves working in a private sector structure, which had the capacity to pay much higher wages and supply more robust benefit packages.

This example shows how the cumulative impacts of larger projects could exacerbate pressures created by worker shortages. In another example, for some communities where multiple projects are already underway, one impact is that restaurants may be unable to remain open because of a shortage of qualified cooks—cooks lost to the high wages of construction camps. The opportunity to realise sales and incomes from the economic boom is lost and the business may not even be able to survive.

frame in the same region can, however, easily overwhelm the local and regional economic structure.

In any community and region, there are typically a range of economic development organisations. These may be general purpose or may be specific to individual groups or sectors. They may be confined to individual communities or they may have a regional mandate. Although such organisations can work within the local economic structure and with local enterprises to take advantage of opportunities provided by individual projects, the cumulative impacts of multiple projects occurring over time means that such organisations may be distracted by the many activities, but they can also be easily overwhelmed by trying to maintain contact with all activities. This is especially challenging because such coordination and planning organisations often operate with a limited budget, and the pressures imposed by the many opportunities can easily move from incremental to negative tipping points.

## 4.7 Issues of Community Resilience

The notion and definition of a *community* have long intrigued and perplexed social science researchers (Sanderson 1938; McClenahan 1946; Hillery 1955, 1972). It is one of those words that is at once familiar to us all, and whose definition seems self-evident, until one tries to define the term to support research on the impacts of change. Although there is much debate in the academic literature about the scope and meaning of *community*, there is some agreement that it reinforces feelings of membership and belonging, and that it generally defines the social and geographic framework within which individuals experience and conduct most of their day-to-day activities (Cater and Jones 1989; Hale 1990; Valentine 2001; Del Casino 2009). Communities serve a range of basic functions, including social participation, mutual support, and a shared sense of identity. It is through their community that individuals organise their daily lives and make sense of the issues and concerns they encounter (Fischer 1982; Halseth and Sullivan 2002).

Historically, two approaches have been taken to defining and exploring the concept of community. The first is a top-down definitional approach. In this case, a boundary (usually administrative) is drawn around the population. A boundary creates forms of inclusion and exclusion that depend on which side of the boundary people are located. These definitions are common for institutional data sources, political jurisdictions, the identification of areas affected by impact assessments, and the like. These administrative communities can vary in scale from a street or neighbourhood to a small town, but it is generally assumed that the residents share certain bonds as a result of sharing a common local environment. Among the problems with such top-down approaches are the fact that they are artificially created and may be very far removed from how people live their lives.

The second approach has been from a bottom-up perspective. In other words, this approach examines the social structures and behaviours of individuals in groups, and the elements that a sense of belonging conveys to members of those groups.

Based on the premise that people are social animals and generally need social contact, researchers have constructed criteria for the reciprocal relationship between the individual and the group and have used them in the study of disruption and change. In such interest-based communities it is not necessary that community members live physically close to each other, but rather that they are brought together by sets of concerns, relationships, or interests. In an increasingly globalised world, and with the increasing use of electronic communication and social media, the formation of interest-based communities has likewise grown.

For both the top-down and bottom-up approaches, classic studies of economic booms and collapses have yielded insights into how stress-induced impacts create wide-ranging changes and transformations that work their way through the social system (see summaries within Ryser and Halseth 2010 and Ryser et al. 2014).

Although useful, both the top-down and the bottom-up approaches have tended towards the creation of checklists for the definition of a community and the identification or tracking of stresses that act upon that community. The challenge with such checklists is that they lose the elements of meaning and context. Community research has identified clearly that it is meaning and context that are absolutely fundamental to the evaluation of cumulative impacts, and especially the impacts that multiple resource industry activities and projects can bring with them.

To better integrate meaning and context, it is more productive to think of community as a process rather than as an entity whose attributes can be listed and tracked (Lewis 1979; Wellman 1987; Halseth 1998). As a process, with the inherent recognition that change is both a normal and a typical part of a functioning community, the next step is to look more closely at those elements driving that process and the changes that affect it. This is where recent interest in social cohesion and social capital by social science researchers can be brought to bear (Lockie et al. 2009; Schirmer 2011). Although much has been written about social cohesion and social capital, leading to debates over certain points, they are at their core rather simple concepts. Social cohesion is about the processes of relationships and interactions that build towards cohesive networks. The products of these processes of interactions are the networks and relationships themselves. These products are the focus of social capital—the trust and bonds that develop from interactions in networks.

Social capital has two additional facets that are important in the consideration of cumulative impacts (Vanclay and Esteves 2011). *Bonding* social capital occurs within groups or in a local setting. A function of bonding social capital is to strengthen networks of information and support so that the group can accomplish things that are beyond the capacity of individuals. In contrast, *bridging* social capital acts as a mechanism for linking between groups, or between local and external organisations and networks.

A cautionary note is required when thinking about social cohesion and social capital. Much of the public policy literature has attempted to translate these terms into positive attributes that support community development. However, it is a general caution for those who are working in community development that social cohesion and social capital processes are neither inherently positive nor negative attributes. Rather, they are simply descriptors of processes within the community.

How those assets of social cohesion and social capital are deployed can determine whether to use a normative positive or negative judgment or evaluation to describe the outcomes.

Cumulative impacts can be significant for social cohesion and social capital. Disruption, change, population loss or gain, and a range of other impacts affect social cohesion and social capital processes and impact how individuals and organisations are able to connect with one another and with a broader range of networks, resources, and expertise. These, in turn, affect the constitution and re-constitution of a community on an ongoing basis. Over time, cumulative impacts on social cohesion and social capital can affect the community's capacity and ability to respond to change or disruption. It is important to note that these cumulative impacts can be positive or negative. Networks can be disrupted and they can take time to realign. Levels of trust and understanding can similarly be disrupted and will also take time to realign. These impacts reflect the delayed shock and response that I raised in Sect. 4.5.2. It is important to note that networks and levels of trust can also be strengthened.

## 4.8 Discussion

The preceding sections have introduced a range of issues that are important if we are to have a conversation that leads towards a more robust and inclusive understanding of the cumulative impacts from resource development projects from a community perspective. One of the key themes running through the chapter is that such an understanding must be both flexible and learning (see Chap. 8). It is simply unrealistic to expect that we will know all of the facets of the local and regional community that will be affected by one or more projects over time. To this point, assessment processes have been locked into a linear and stepwise process that leads towards some definitive impact statement that runs counter to the way the world works, and counter to our emerging understanding of that world. Such linear approaches are of limited value. Circumstances will be subject to change over time, and this must be an expected and integral part of any cumulative impact assessment and monitoring. Any approach must allow for recursive learning.

A second theme running through the chapter is the challenge that comes when impacts are cast only in negative terms. I argue that assigning such normative labels may not only be limiting our ability to understand short-, medium-, and long-term outcomes, but that it may preclude dialogue about the opportunities that come with any challenge. In agreement with this book's call for a more integrative approach to understanding regional cumulative impacts, it is only with a more holistic understanding of both the challenges and the opportunities that we can see how projects or proposals fit with community and regional aspirations.

A third theme running through the chapter concerns the need to recognise some inherent limits to cumulative impact assessment processes. A range of tensions act to limit our ability, and even our desire, to evaluate cumulative impacts into the near and distant future. Most people understand the imperative of achieving sustainable and resilient communities, economies, and ecosystems, but there are also the realities of needing to put food on the table and pay bills each day! Even where linear and stepwise assessment processes are in place, tensions between organisations can limit the effectiveness of any framework (Noble et al. 2013). There is also the challenge of causality, which is an especially limiting factor in how to approach cumulative impact assessments and debates. The response from a community perspective is the need to consider likely connections and contributions, and leave matters of the legal attribution of blame to the legal system.

This chapter also introduced the importance of context and scale in considering the cumulative impacts on a community from multiple large or small, and past or present, resource development projects. One important notion of context is that different places will be affected differently. These impacts may have shorter or longer durations depending upon the type of activities and the capacity of the communities to be resilient in the face of those activities. Such impacts may also differ among the different projects that are operating within a region. Dealing with multiple and asynchronous *waves* of impacts further complicates the delineation of these impacts and our ability to respond (Ryser et al. 2014). In addition, different communities will quite naturally be differentially equipped in terms of their experience and resilience with respect to changes and natural resource development projects. In my consideration of geographic scale, I noted the importance of considering multiple scales, ranging from the individual to the region. Again, different capacities and resilience, together with different support structures that must be set in place to assist with coping and resilience, will affect how cumulative impacts unfold and are addressed. The need for a more integrative and regional approach to cumulative impact assessment finds much support when impacts are considered from a community perspective.

Dana et al. (2009) studied oil and gas development in Canada's NWT. They found that even though residents recognised the short-term economic advantages of such projects, they had great concerns "about the long-term impact on the environment that currently ensures their livelihood" (Dana et al. 2009, p. 94). These findings are reinforced by a recent study by Groth and Vogt (2014), who examined resident perceptions related to the social, environmental, or economic impacts of wind farm development in a selection of rural counties in the USA. They found an increasing importance of non-economic factors in the decisions of residents about whether to support or oppose additional wind farm development.

At the community level, my consideration of cumulative impacts focused upon two frameworks. The first involved a set of four infrastructures that are critical for local capacity and the resilience to cope with both opportunities and challenges. Physical infrastructure, human infrastructure, community infrastructure, and economic infrastructure not only need to be assessed in advance of any significant new economic initiatives, but they must also be supported throughout the development process even if the fiscal resources (e.g. tax revenues) from those new industries may not yet be available to build or sustain the infrastructure. The second critical framework involves the notions of social cohesion and social capital, which are essential for maintaining resilient and sustainable communities. The challenge for many rural and small town communities is that, by definition, they have small populations that have correspondingly small human resources, and that are easily disrupted by processes that include the mobility of people into and out of the community. Again, it is critical that assessments of cumulative impacts not only account for the status of social cohesion and social capital prior to new economic development initiatives, but are also available to track the impacts on these factors and support the renewal of both during the transitions from project planning to construction, and from construction into operations.

In discussions about the impacts of industrial resource development projects on communities, there continues to be little attention devoted to the fundamental challenges that come when the "magnitude of external boom-bust forces may be so great as to overwhelm even the best-prepared communities" (Gramling and Freudenburg 1990, p. 541). In a review of policy and regulatory decision-making processes, Barth (2013) emphasised that significant assumptions continue to be made around the degree of economic prosperity that unconventional gas development may bring to state and local economies. Her results suggest that to date, at least in the USA, the benefits are mixed.

The topics in this chapter agree with those outlined by Noble (2010): to understand cumulative impacts, the approach must be oriented towards the future, based on a range of alternatives, integrative, adaptable, multi-scaled, ecosystem-based, multi-sector, and multi-scale. A community-based, integrative, and regionally inclusive cumulative impact assessment must have a temporal orientation that extends from the past to the present, and into the future. It must also be flexible and recursive, because resource development projects change through their planning and implementation stages. In terms of being a learning process, any approach must be comprehensive and integrated with a wide range of issues and information sources on those issues, and it must be open to adaptation over time. Similarly, the approach must recognise differences in geographic and temporal scales, and how these different scales interact with one another. In summary, these approaches must be systems based, reach across broad community development sectors, and be based on a complex understanding of the multiple levels of impacts that will occur as impacts unfold in both expected and unexpected ways over time.

## 4.9 Conclusions

In this chapter, I have explored how a community perspective can lead to a more comprehensive understanding of the integrative regional cumulative impacts from resource development projects. As is the case for environmental and health impacts, any impacts on a community will be complex and interwoven, and will range across geographic and temporal scales. Building upon some of the foundational elements that guide research on community development and the changes it causes for rural and small town communities, I propose that more attention must be paid to the four basic community infrastructures and to the processes that affect social cohesion and social capital. These are useful starting points for a community-based conversation leading towards an integrative regional cumulative impact assessment approach.

No matter the approach, it is important to remember that although there will be various and cumulative impacts from resource development projects of any size, there will also be (potentially cumulative) impacts when such projects do not take place. In working with communities and regions, communities will need to be ready whether or not projects happen.

## References


Arce-Gomez, A., J. Donovan, and R. Bedggood. 2015. Social impact assessments: Developing a consolidated conceptual framework. *Environmental Impact Assessment Review* 50: 85–94.

Australian House of Representatives. 2013. *Cancer of the bush or salvation for our cities? Fly-in, fly-out and drive-in, drive-out workforce practices in regional Australia*. Canberra, Australia: Standing Committee on Regional Australia, Parliament of the Commonwealth of Australia.

Bakke, B., B. Ulvestad, P. Stewart, and W. Eduard. 2004. Cumulative exposure to dust and gases as determinants of lung function decline in tunnel construction workers. *Occupational and Environmental Medicine* 61: 262–269.

Barnes, T. 1996. External shocks: Regional implications of an open staple economy. In *Canada and the global economy: The geography of structural and technological change*, ed. J. Britton, 48–68. Montreal: McGill Queen's University Press.

Barth, J.M. 2013. The economic impact of shale gas development on state and local economies: benefits, costs, and uncertainties. *New Solutions* 23: 85–101.

Bone, R. 1992. Mega-projects in northern development. In *The geography of the Canadian North: Issues and challenges*, ed. R. Bone, 135–156. Toronto: Oxford University Press.

Boschma, R., and K. Frenken. 2006. Why is economic geography not an evolutionary science? Toward an evolutionary economic geography. *Journal of Economic Geography* 6: 272–302.

Boschma, R., and R. Martin (eds.). 2010. *Handbook of evolutionary economic geography*. Cheltenham: Edward Elgar.

Brown, R.B., S.F. Dorius, and R.S. Krannich. 2005. The boom-bust-recovery cycle: Dynamics of change in community satisfaction and social integration in Delta, Utah. *Rural Sociology* 70: 28–49.

Cater, J., and T. Jones. 1989. *Social geography*. London: Edward Arnold.

Cayo, D. 2014. Without agreed-on facts, environmental reviews are mired in dysfunction. *Vancouver Sun*, September 19. http://www.vancouversun.com/Cayo+Without+agreed+facts+environmental+reviews+mired+dysfunction+with+video/10219427/story.html#ixzz3DoR5hIFL. Accessed 19 Sep 2014.

Dana, L.P., R.B. Anderson, and A. Meis-Mason. 2009. A study of the impact of oil and gas development on the Dene First Nations of the Sahtu (Great Bear Lake) region of the Northwest Territories (NWT). *Journal of Enterprising Communities: People Places in the Global Economy* 3: 94–117.

Del Casino, V.J. 2009. *Social geography*. Oxford: Wiley-Blackwell.

Disability Alliance BC. 2014. BC Disability Benefits Help Sheet 13: Rate amounts for persons with disabilities (PWD) and persons with persistent and multiple barriers to employment (PPMB) benefits. http://www.disabilityalliancebc.org/docs/hs13.pdf?LanguageID=EN-US. Accessed 1 Apr 2015.

Drache, D. 1991. Harold Innis and Canadian capitalist development. In *Perspectives on Canadian economic development: Class, staples, gender, and elites*, ed. G. Laxer, 22–49. Don Mills: Oxford University Press Canada.

Drahokoupil, J. 2012. Beyond lock-in versus evolution, towards punctuated co-evolution: On Ron Martin's 'Re-thinking regional path dependence'. *International Journal of Urban and Regional Research* 36: 166–171.

Emery, M., and C. Flora. 2006. Spiraling-up: Mapping community transformation with community capitals framework. *Community Development* 37: 19–35.

England, J., and S. Albrecht. 1984. Boomtowns and social disruption. *Rural Sociology* 49: 230–246.

Esteves, A., and F. Vanclay. 2009. Social development needs analysis as a tool for SIA to guide corporate-community investment: Applications in the minerals industry. *Environmental Impact Assessment Review* 29: 137–145.

Fidler, C., and B. Noble. 2012. Advancing strategic environmental assessment in the offshore oil and gas sector: Lessons from Norway, Canada, and the United Kingdom. *Environmental Impact Assessment Review* 34: 12–21.

Fischer, C.S. 1982. *To dwell among friends—personal networks in town and city*. Chicago: University of Chicago Press.

Franks, D.M., D. Brereton, and C.J. Moran. 2011. Cumulative social impacts. In *New directions in social impact assessment: Conceptual and methodological advances*, ed. F. Vanclay and A.M. Esteves, 202–220. Cheltenham: Edward Elgar Publishing.

Freudenburg, W. 1981. Women and men in an energy boomtown: Adjustment, alienation, and adaptation. *Rural Sociology* 46: 220–244.

Freudenburg, W.R. 1984. Boomtown's youth: The differential impacts of rapid community growth on adolescents and adults. *American Sociological Review* 49: 697–705.

Gachechiladze-Bozhesku, M., and T. Fischer. 2012. Benefits of and barriers to SEA follow-up: Theory and practice. *Environmental Impact Assessment Review* 34: 22–30.

Gartrell, J.W., and H. Krahn. 1983. *Housing change and perceived quality in a Northern Alberta Boomtown*. Edmonton: Department of Sociology, University of Alberta.

Government of Alberta. 2006. *Investing in our future: Responding to the rapid growth of oil sands development*. Edmonton: Government of Alberta.

Grafton, Q., and N.R. Lambie. 2014. *Australia's experience in developing an LNG export industry*. Vancouver: Asia Pacific Foundation of Canada in Partnership with the Australian Pacific Economic Cooperation Committee.

Gramling, R., and W.R. Freudenburg. 1990. A closer look at "local control": Communities, commodities and the collapse of the coast. *Rural Sociology* 55: 541–558.

Groth, T.M., and C.A. Vogt. 2014. Rural wind farm development: Social, environmental and economic features important to local residents. *Renewable Energy* 63: 1–8.

Hale, S.M. 1990. *Controversies in sociology—a Canadian introduction*. Toronto: Copp Clark Pitman Ltd.

Haley, B. 2011. From staples trap to carbon trap: Canada's peculiar form of carbon lock-in. *Studies in Political Economy* 88: 97–132.

Halseth, G. 1998. *Cottage country in transition: A social geography of change and contention in the rural-recreational countryside*. Montreal: McGill-Queen's University Press.

Halseth, G., S. Markey, and D. Bruce. 2010. *The next rural economies: Constructing rural place in global economies*. Oxfordshire: CABI Publishing.

Halseth, G., and L. Ryser. 2006. Trends in service delivery: Examples from rural and small town Canada, 1998 to 2005. *Journal of Rural and Community Development* 1: 69–90.

Halseth, G., and L. Sullivan. 2002. *Building community in an instant town: A social geography of Mackenzie and Tumbler Ridge, BC*. Prince George: University of Northern British Columbia Press.

Hanlon, N., and G. Halseth. 2005. The greying of resource communities in northern British Columbia: Implications for health care delivery in already-underserviced communities. *The Canadian Geographer* 49: 1–24.

Hanlon, N., M. Skinner, A. Joseph, L. Ryser, and G. Halseth. 2014. Place integration through efforts to support healthy aging in resource frontier communities: The role of voluntary sector leadership. *Health and Place* 29: 132–139.

Harriman, J., and B. Noble. 2008. Characterizing project and strategic approaches to regional cumulative effects assessment in Canada. *Journal of Environmental Assessment Policy and Management* 10: 25–50.
Harvey, D. 1990. *The condition of postmodernity: An inquiry into the origins of cultural change*. Oxford: Blackwell.
———. 2005. *A brief history of neoliberalism*. Oxford: Oxford University Press.
Hillery Jr., G.A. 1955. Definitions of community: Areas of agreement. *Rural Sociology* 20: 111–123.
———. 1972. Selected issues in community theory. *Rural Sociology* 37: 534–552.
Hudson, R. 2005. Region and place: Devolved regional government and regional economic success? *Progress in Human Geography* 29: 618–625.
Innis, H. 1933. *Problems of staple production in Canada*. Toronto: Ryerson Press.
Kinnear, S., Z. Kabir, J. Mann, and L. Bricknell. 2013. The need to measure and manage the cumulative impacts of resource development on public health: An Australian perspective. In *Current topics in public health*, ed. A. Rodriguez-Morales, 125–144. Rijeka: InTech Publishers.
Kretzmann, J.P., and J.L. McKnight. 1993. *Building communities from the inside out: A path toward finding and mobilizing a community's assets*. Chicago: ACTA Publications.
Krzyzanowski, J., and P.L. Almuedo. 2010. *Cumulative impacts of natural resource development on ecosystems and wildlife: An annotated bibliography for British Columbia. Prepared for the BC Ministry of Forests and Range, Forest Investment Account-Forest Science Program*. Kamloops: Forrex Forum for Research and Extension in Natural Resources.
Lewis, G.J. 1979. *Rural communities: Problems in modern geography*. London: David and Charles Ltd.
Lo, J., and G. Halseth. 2009. The practice of principles: An examination of CED Groups in Vancouver, BC. *Community Development Journal* 44: 80–110.
Lockie, S., M. Franettovich, V. Petkova-Timmer, J. Rolfe, and G. Ivanova. 2009. Coal mining and the resource community cycle: A longitudinal assessment of the social impacts of the Coppabella coal mine. *Environmental Impact Assessment Review* 29: 330–339.
Loxton, E.A., J. Schirmer, and P. Kanowski. 2013. Exploring the social dimensions and complexity of cumulative impacts: A case study of forest policy changes in Western Australia. *Impact Assessment and Project Appraisal* 31: 52–63.
MacLeod, G. 2001. New regionalism reconsidered: Globalization and the remaking of political economic space. *International Journal of Urban and Regional Research* 25: 804–829.
Malamud, G.W. 1984. *Boomtown communities*. New York: Van Nostrand Reinhold.
Markey, S., G. Halseth, and D. Manson. 2008. Closing the implementation gap: A framework for incorporating the context of place in economic development planning. *Local Environment* 13: 337–351.
———. 2012. *Investing in place: Economic renewal in Northern British Columbia*. Vancouver: UBC Press.
Markey, S., and K. Heisler. 2011. Getting a fair share: Regional development in a rapid boom-bust rural setting. *Canadian Journal of Regional Science* XXXIII: 49–62.
Martin, R. 2000. Institutional approaches in economic geography. In *A companion to economic geography*, ed. E. Sheppard and T. Barnes, 77–94. Oxford: Blackwell Publishers.
———. 2010. Roepke lecture in economic geography—rethinking regional path dependence: Beyond lock-in to evolution. *Economic Geography* 86: 1–27.
Massey, D. 1995. *Spatial divisions of labour*. London: Macmillan.
McClenahan, B.A. 1946. The communality: The urban substitute for the traditional community. *Sociology and Social Research* 30: 264–274.
McCreary, T. 2013. Mining Aboriginal success: The politics of difference in continuing education for industry needs. *Canadian Geographer* 57: 280–288.
Michell, G., and P. McManus. 2013. Engaging communities for success: Social impact assessment and social licence to operate at Northparkes Mines, NSW. *Australian Geographer* 44: 435–459.
Moran, C.J., D.M. Franks, and L.J. Sonter. 2013. Using the multiple capitals framework to connect indicators of regional cumulative impacts of mining and pastoralism in the Murray Darling Basin, Australia. *Resources Policy* 38: 733–744.

Morris, R. 2012. *Scoping study: Impact of fly-in fly-out/drive-in drive-out work practices on local government*. Sydney: Australian Centre of Excellence for Local Government, University of Technology.

Nielsen, S., and M. Elle. 2000. Assessing the potential for change in urban infrastructure systems. *Environmental Impact Assessment Review* 20: 403–412.

Nitschke, C.R. 2008. The cumulative effects of resource development on biodiversity and ecological integrity in the Peace-Moberly region of Northeast British Columbia, Canada. *Biodiversity and Conservation* 17: 1715–1740.

Noble, B. 2010. *Cumulative environmental effects and the Tyranny of small decisions: Towards meaningful cumulative effects assessment and management*. Prince George: Natural Resources and Environmental Studies Institute, University of Northern British Columbia. Occasional Paper No. 8.

Noble, B.F. 2009. Promise and dismay: The state of strategic environmental assessment systems and practices in Canada. *Environmental Impact Assessment Review* 29: 66–75.

Noble, B.F., J.S. Skwaruk, and R.J. Patrick. 2013. Toward cumulative effects assessment and management in the Athabasca watershed, Alberta, Canada. *The Canadian Geographer* 58: 315–328.

Oosterlynck, S. 2012. Path dependence: A political economy perspective. *International Journal of Urban and Regional Research* 36: 158–165.

Owen, J.R., and D. Kemp. 2013. Social licence and mining: A critical perspective. *Resources Policy* 38: 29–35.

Page, S. 2006. Path dependence. *Quarterly Journal of Political Science* 1: 87–115.

Palen, W.J., T.D. Sisk, M.E. Ryan, J.L. Arvai, M. Jaccard, A.K. Salomon, T. Homer-Dixon, and K.P. Lertzman. 2014. Comment: Consider the global impacts of oil pipelines. *Nature* 510: 465–467.

Perry, S. 2013. Using ethnography to monitor the community health implications of onshore unconventional oil and gas developments: Examples from Pennsylvania's Marcellus Shale. *New Solutions* 23: 33–53.

Peterson, E.B., Y.H. Chan, N.M. Peterson, G.A. Constable, R.B. Caton, C.S. Davis, R.R. Wallace, and G.A. Yarronton. 1987. *Cumulative effects assessment in Canada: An agenda for action and research*. Hull: Canadian Environmental Assessment Research Council.

Plate, E., M. Foy, and R. Krehbiel. 2009. *Best practices for First Nation involvement in environmental assessment reviews of development projects in British Columbia*. West Vancouver: New Relationship Trust.

Polèse, M. 1999. From regional development to local development: On the life, death, and rebirth(?) of regional science as a policy relevant science. *Canadian Journal of Regional Science* XXII: 299–314.

Porter, M. 2000. Location, competition, and economic development: Local clusters in a global economy. *Economic Development Quarterly* 14: 15–34.

Prno, J. 2013. An analysis of factors leading to the establishment of a social licence to operate in the mining industry. *Resources Policy* 38: 577–590.

Reimer, B. 2010. Space to place: Bridging the gap. In *The next rural economies*, ed. G. Halseth, S. Markey, and D. Bruce, 263–274. Wallingford: CABI International.

Ryser, L., and G. Halseth. 2010. Rural economic development: A review of the literature from industrialized economies. *Geography Compass* 4: 510–531.

———. 2011. Housing costs in an oil and gas boom town: Issues for low-income senior women living alone. *Journal of Housing for the Elderly* 25: 306–325.

———. 2014. On the edge in rural Canada: The changing capacity and role of the voluntary sector. *Canadian Journal of Nonprofit and Social Economy Research* 5: 41–56.

Ryser, L., S. Markey, D. Manson, J. Schwamborn, and G. Halseth. 2014. From boom and bust to regional waves: Development patterns in the Peace River Region, British Columbia. *Journal of Rural and Community Development* 9: 87–111.

Sanderson, D. 1938. Criteria of community formation. *Rural Sociology* 3: 371–384.

Savoie, D. 1997. *Rethinking Canada's regional development policy: An Atlantic perspective*. Ottawa: Canadian Institute for Research on Regional Development.

Schirmer, J. 2011. Scaling up: Assessing social impacts at the macro-scale. *Environmental Impact Assessment Review* 31: 382–391.

Scott, M. 2004. Building institutional capacity in rural Northern Ireland: The role of partnership governance in the LEADER II programme. *Journal of Rural Studies* 20: 49–59.

Shandro, J.A., M.M. Veiga, J. Shoveller, M. Scoble, and M. Koehoorn. 2011. Perspectives on community health issues and the mining boom-bust cycle. *Resources Policy* 36: 178–186.

Storper, M. 1999. The resurgence of regional economics: Ten years later. In *The new industrial geography: Regions, regulation and Institutions*, ed. T.J. Barnes and M.S. Gertler, 23–53. New York: Routledge.

Sullivan, L., L. Ryser, and G. Halseth. 2015. Recognizing change, recognizing rural: The new rural economy and towards a new model of rural services. *Journal of Rural and Community Development* 9: 219–245.

Valentine, G. 2001. Community. In *Social geographies: Society and space*, 105–138. Essex: Pearson Education Ltd.

Vanclay, F., and A.M. Esteves (eds.). 2011. *New directions in social impact assessment: Conceptual and methodological advances*. Northampton: Edward Elgar Publishing.

Van Hinte, T., T. Gunton, and J. Day. 2007. Evaluation of the assessment process for major projects: A case study of oil and gas pipelines in Canada. *Impact Assessment and Project Appraisal* 25: 123–137.

Watkins, M. 1982. The Innis tradition in Canadian political economy. *Canadian Journal of Political Science and Social Theory* 6: 12–34.

Wellman, B. 1987. *The community question re-evaluated*. Toronto: University of Toronto, Centre for Urban and Community Studies. Research Paper No. 165.

Wong, C., and W. Ho. 2015. Roles of social impact assessment practitioners. *Environmental Impact Assessment Review* 50: 124–133.

# Chapter 5
# Cumulative Determinants of Health Impacts in Rural, Remote, and Resource-Dependent Communities

**Margot W. Parkes**

*With contributions by* **Henry G. Harder, Dawn Hemingway, Martha L. P. MacLeod, Pouyan Mahboubi, Indrani Margolin, and Cathy Ulrich**

## 5.1 Introduction

For those who are working in health research, policy, and practice, the phrase *cumulative health impacts* poses interesting challenges. On the one hand, the need to understand the combined impact of multiple stressors on human health and well-being is a natural progression and application of our current knowledge about human health. On the other hand, anyone trying to grasp cumulative health impacts must confront the questions and debates involved in defining each word separately (i.e. cumulative, health, and impacts), as well as the complexity of addressing these concepts together. Many of these questions have been addressed in earlier chapters, including the following: What constitutes an impact (Chaps. 1–4)? What do we mean by cumulative (Sect. 2.2 in Chap. 2)? How do we take into account the combined impacts of multiple drivers of changes over both time and space (Chap. 2)? The perspectives on health and well-being introduced in this chapter also demand consideration of another fundamental question: What do we mean by *health*? Many discussions of the influence of resource development on health tend to focus on the presence or absence of *disease* attributable to a specific hazard or exposure pathway. A different conversation about cumulative impacts occurs if health impacts are framed in relation to the cascade of complex interactions that determine health and well-being in the long term.

This chapter is premised on the idea that cumulative health impacts need to be understood as part of ongoing efforts to address the combined social and environmental *determinants of health* (Schulz and Northbridge 2004; Marmot 2005; Reading

M.W. Parkes (✉)
The University of Northern British Columbia, Prince George, BC, Canada V2N 4Z9
e-mail: Margot.Parkes@unbc.ca


M.P. Gillingham et al. (eds.), *The Integration Imperative*,
DOI 10.1007/978-3-319-22123-6_5

and Wien 2009; Parkes et al. 2010; CPHA 2015). This demands integrative approaches to impact assessment that explicitly engage with the rapid social and ecological changes occurring in the twenty-first century (Briggs 2008; Harris-Roxas and Harris 2011) and that are also informed by an understanding of the cumulative, environmental, and community dynamics introduced in Chaps. 1–4.

The chapter begins with an overview of current approaches to understanding and responding to health impacts and identifies a spectrum of prospective, reactive, and integrative approaches that inform Canadian and global approaches to assessing health impact. It then highlights key considerations relevant to understanding the health impacts of resource development in rural, remote, and resource-dependent communities by drawing on exemplars and experiences within northern BC, Canada. These examples identify a range of specific health-related issues including health sector priorities in relation to resource development, and highlight new insights gained by focusing on the experiences of specific populations (including older residents and Aboriginal[1] peoples) and considering psychological as well as physical impacts of resource development. Related issues are also explored in Chap. 6 (see Vignettes 6.5–6.7). The chapter concludes by discussing the examples from northern BC in relation to a cascade of direct and indirect impacts on social and environmental determinants of health.

The interface between impact assessment and the determinants of health is an important challenge in its own right, but is especially important in the context of the environmental, community, and health impacts of resource development. Combining an awareness of the cumulative impacts of resource development with a consideration of the determinants of health from a variety of perspectives—including mental health, Aboriginal health, and social and environmental justice concerns—provides a more integrated view of the *cumulative determinants of health impacts*. Such efforts build and extend on our current understanding of the social determinants of health—often described as the *causes of the causes* of health inequalities (Marmot 2005; Raphael 2006; CSDH 2008; Reading and Wien 2009; Mikkonen and Raphael 2010)—and also respond to calls to (re)integrate ecological and ecosystem perspectives into our understanding of the interrelated factors that determine health and well-being (Waltner-Toews 2004; Corvalan et al. 2005; Webb et al. 2010; CPHA 2015).

## 5.2 Health Impact Assessment Challenges: Prospective, Reactive, and Integrative Approaches

Any contemporary consideration of cumulative health impacts needs to be situated within the larger spectrum of research, policy, and practice concerned with health impacts, assessments, and determinants of health. The background and context

[1] As we noted in Sect. 1.7, the term *Aboriginal* is used to refer inclusively to First Nations, Métis, and Inuit peoples in Canada. Depending on the context, we have also used *First Nation* (in Canadian political contexts) and *Indigenous* (generally in international contexts).

presented in this section are intended to flag some key developments relating to the assessment of health impacts, prior to considering cumulative determinants of health impacts in more detail. In order to focus the discussion in this health-oriented chapter, working definitions are helpful. Throughout the chapter *health* refers to the physical, social, and psychological well-being of an individual, or sometimes of a community, and does not generally focus on a specific health outcome or disease. In keeping with Box 1.1, an *impact* represents the consequence of some *effect* (change) caused by natural resource development, but here, the focus is on the consequences for health. Finally, *assessment* refers to any way of understanding health impacts sufficiently well that it becomes possible to propose an intervention that would mitigate the impacts.

From a health perspective, discussion of cumulative impacts raises a range of terminological and definitional challenges. Even in isolation, health impact assessment (HIA) has a variety of definitions and applications that depend on the context in which an assessment is initiated (Morgan 2003; Harris-Roxas and Harris 2011). As the demand to consider health impacts and their cumulative dynamics continues to grow, a diverse set of terms and meanings have emerged that span a wide range of literatures. These include the following:

- *Cumulative environmental hazards index*: This approach proposes an index that draws on chemical data and social vulnerability as well as a health index (Huang and London 2012).
- *Cumulative environmental justice impact assessment*: Kreig and Faber (2004) propose a composite measure of social conditions and cumulative ecological hazards in order to identify and compare the disproportionate impacts of ecologically hazardous sites and facilities on low-income communities and communities of colour.
- *Cumulative risk assessment*: This approach responds to the increasing pressure for regulatory agencies to consider public health concerns related to multiple pollutant exposures, pathways, and vulnerable populations, but there is widespread recognition that cumulative risk assessment can yield inexact and uncertain results (Briggs 2008; Lewis et al. 2011).
- *Environmental health impact assessment*: Initially proposed by Fehr (1999) as a framework with which to understand the impacts of a waste disposal facility, this approach was subsequently developed by McCarthy et al. (2002) to focus on quantification of environmental health effects.
- *Generalized health impact*: Using Canadian Community Health Survey and Census data, this approach models cumulative effects of social exposures on different health outcomes, by comparing "disease-specific" and "generalised health-impact" models to gauge the negative health effects of socioeconomic position (White et al. 2013).
- *Integrated environmental health impact assessment*: This approach provides a nested framework for an integrated, environmental HIA, making comparisons and contrasts with risk assessment, standard HIA, and comparative risk assessment (Briggs 2008).

Although not comprehensive, these examples provide an important reminder of the spectrum of different languages and terminologies that have emerged in relation to health risks, impacts, and cumulative considerations, and of the resulting potential for overlaps, confusion, and misunderstanding around terms and related discourses (see also Corburn 2002). Given the range of practical and scholarly histories and lineages involved, this diversity should not be seen as a problem to be overcome or corrected through unification. Instead, the diversity of language should be acknowledged as an expression of complexity and a caution against over-simplification in a context that warrants comprehensive responses. New terminology will continue to emerge, not least to reflect increased understanding of the determinants of health impacts, as exemplified by the interplay between SIA, socioeconomic disparities, and health equity impact assessment (see Box 4.1; Snyder et al. 2012; White et al. 2013; Povall et al. 2014). At the same time, cautions have arisen about developing new terminology unnecessarily. In their review of the capacity of existing HIA methodologies to adequately address health equity impacts, Povall et al. (2014) provide recommendations to strengthen attention to health equity and inequities in existing HIA processes, but do not recommend adopting new terminology to do so.

Against this backdrop of ongoing developments, the field of HIA remains the most developed overarching area of research and application, and offers an important and informative starting point for considering cumulative health impacts, especially if health is to be considered in the context of wider environmental and community dynamics. Detailed accounts of the strengths and weaknesses of HIA frameworks can be found elsewhere (Morgan 2003, 2011; Wismar et al. 2007; Mindell et al. 2008; Lee et al. 2013). These authors highlight not only the variation in frameworks, but also that the definition of HIA will vary depending on the context in which it is implemented.

Informed by this book's specific focus on resource development contexts, this chapter identifies and distinguishes three different approaches to assessing health impacts: first, proactive approaches which are often independently driven, stand-alone health impact assessments; second, *reactive* approaches whereby legislative and proponent-driven approaches address health concerns within other assessment processes; and third, *integrative* approaches that are emerging in response to the specific challenges of intensive resource development and are described in more detail. The distinctions among these three approaches are especially relevant to Canada and northern BC, but will also be relevant in other jurisdictions, as they provide points of reference to consider as part of efforts to develop more cumulative and integrative approaches to assessing health impacts.

### 5.2.1 Health Impact Assessment as a Proactive, Discrete Process

The first approach considered here is a proactive approach that is most closely associated with HIA as a discrete, *stand-alone* process. St-Pierre and Mendell (2012, p. 275) provide a definition of HIA that highlights the proactive potential of HIA:

"Health Impact Assessment is a five-step, prospective process to evaluate the potential health effects of a policy proposal from outside the health sector". Several decades of experience, practice, and literature have led to a broad consensus around the procedural aspects of HIA as a discrete process conducted in a series of stages (see, for example, Wismar et al. 2007; Wernham 2011; Harris-Roxas et al. 2012; WHO 2014) including:

- Screening (establishing whether an HIA is warranted);
- Scoping (planning the HIA, and determining the required information and resources);
- Appraisal or assessment of impacts (undertaking the HIA by collecting and analysing information);
- Reporting (sharing the results with interested parties, including via a written report); and
- Evaluation, monitoring, and follow-up (determining the effects of the HIA and of any proposed policy or intervention).

In the 1990s, a researcher, policy maker, or educator who was seeking to understand HIA as a proactive approach for understanding the consequences of resource development may have turned to Canada as a point of reference. Indeed, while HIA was developing as a field, Canada provided notable international leadership in understanding HIA as a tool for engaging with intersectoral concerns, and as a basis for guiding our approaches to understanding the prospective and retrospective health impacts of changes to our social and physical environments (Frankish et al. 1996; Health Canada 1999, 2004). Despite these early efforts, Canada has not stayed at the forefront of developments in HIA, not least due to an increased orientation on the kinds of reactive assessments that will be discussed in the next section. The variations in, and barriers to progress on, HIA in the Canadian context warrant ongoing critical attention and are especially important when taking cumulative dynamics of resource developments into consideration (see Noble and Bronson 2005, 2006; Mendell 2011; Shandro et al. 2011).

Fortunately, international insights arising from the application of HIA as a discrete, stand-alone process are consolidating an understanding of "state of the art" HIA (Harris-Roxas et al. 2012, p. 43) including its application as a proactive and systematic approach to identifying the health impacts of resource development (Utzinger et al. 2005; Winkler et al. 2012). Proactive approaches to HIA have also informed the increased attention to integrating environmental and health impact assessment in the USA (Dannenberg et al. 2006; Bhatia 2007; Bhatia and Wernham 2008), along with work in Europe, Australia, and New Zealand that is focused on HIA as a tool for intersectoral actions and for considering the health implications for all policy sectors; for an indication of the scope of the available guides and frameworks, see Harris-Roxas and Harris (2011), Lee et al. (2013), WHO (2014), and the World Health Organization guides (http://www.who.int/hia/about/guides/en). Despite the growing international experience the proactive, discrete approach to HIA is often seen as external to regulatory contexts, and as something that is either a focus of independent research or an ad hoc initiative to understand particular issues or contexts. In parallel, another approach to assessing the health implications of resource development has evolved, in which health is framed as merely one impact of (potential) interest within the context of project-based environmental impact assessment legislation and frameworks.

### 5.2.2 *Assessment of Health Impacts as a Reactive, Legislatively Driven Process*

A more reactive approach to assessing health impacts occurs when the assessment is triggered in response to a proposal, and subsequently operates as part of an existing environmental assessment process, with an emphasis on the stressor- and project-based approaches discussed in Chap. 2 (see also Vignette 6.8). In Canada, these processes tend to be legislatively determined and driven in response to the submission of an application by an industrial proponent of a project. Many would argue that such approaches rarely meet the technical or procedural requirements of HIA in the sense described in the previous section, or the "state of the art" HIA described by Harris-Roxas et al. (2012). Even so, these approaches are presented as mechanisms that can be used to assess the health implications of resource development and are sometimes referred to as health impact assessments even if that terminology is not technically accurate.

Since the mid-1990s, many efforts to assess health impacts have been conducted within the framework of the *Canadian Environmental Assessment Act* (Government of Canada 2012; see also Chaps. 1, 2, 7, and 8). Although the *Act* is intended to protect the environment and human health, it is not essential that an EA consider human health. Rather, there are several discretionary points in the process defined by the *Act* when health considerations may be triggered and therefore included in the assessment. Box 5.1 (supported by Fig. 5.1) provides an overview of opportunities to consider health within the *Act* process.

The descriptions in Box 5.1 and Fig. 5.1 apply to most projects for which the Canadian Environmental Assessment Agency (CEAA) is the responsible authority. However, under the *Canadian Environmental Assessment Act,* projects may also fall under the responsible authority of the National Energy Board or the Canadian Nuclear Safety Commission (Government of Canada 2012). Such projects may flow through the system slightly differently, especially with regard to assessment by the responsible authority and to matters relating to substitution (i.e. a single assessment rather than separate assessments when both federal and provincial assessments are required) or equivalency (i.e. exemption from a federal assessment when a provincial assessment will meet the same criteria). Box 5.1 and Fig. 5.1 should therefore not be interpreted as universal, even in the Canadian context, but instead provide points of reference for understanding how health impacts are potentially considered when the CEAA is the responsible authority.

Box 5.1 and Fig. 5.1 help to underscore the specific limitations and constraints faced when trying to understand health impacts as part of a reactive, legislative approach. An obvious dilemma with framing HIA within the context of the

**Box 5.1 An Overview of Opportunities Within the *Canadian Environmental Assessment Act* to Consider Potential Impacts on Human Health (Pouyan Mahboubi)**

In Canada, as in many countries (e.g. Australia and New Zealand), attempts to assess the impacts of proposed policies and projects on human health are channelled through an EA process. The principal mandate of the *Canadian Environmental Assessment Act* is to protect both the environment and human health (Government of Canada 2012, Section 4.2 [s4.2]). The effects on eco-systems are relatively well defined in the *Canadian Environmental Assessment Act*, which references several other Acts of Parliament, including the *Fisheries Act* [s5.1a(i)], (Government of Canada 1985), the *Migratory Birds Convention Act* [s5.1a(iii)] (Government of Canada 1994) and the *Species at Risk Act* [s5.1a(ii)] (Government of Canada 2002). These serve to set the tone for the EA process, largely by focusing it on ecosystem studies. Human health, however, is not well defined and there are few places in the *Canadian Environmental Assessment Act* where health concerns are specifically or systematically included. Figure 5.1 presents an overview of the *Canadian Environmental Assessment Act* based on six major steps, and identifies specific challenges and opportunities for the inclusion of human health considerations, where applicable. The steps (and the relevant sections of the *Act*) are as follows:

*Step 1.* A description of the proposed project is provided by the proponent to the CEAA [s8]. The description is then made available to the public [s9a-b].

*Step 2.* The CEAA screens the proposed project to determine whether an EA is required [s10]. Public input at this step [s9c] over potential health concerns may contribute to the triggering of an EA.

*Step 3.* When an EA is required, the Minister then delegates responsibility for overseeing the process to either a review panel [s38.1–2] or a specific authority: the National Energy Board, the Canadian Nuclear Safety Commission, the CEAA or another comparable federal or provincial authority [s14.4a-d and s15a-d].

*Step 4.* The selected authority must define the scope of factors to be considered in the EA [s19.2]. The process depicted in Fig. 5.1 is applicable when the CEAA is the responsible authority. The typical factors taken into account are changes caused to ecosystem components such as fish and wildlife [s5.1a(i)-(iii)]. Comments from the public can potentially influence the factors to be considered at this step in the process [s19.1c] if they are in accordance with the *Act*. The impacts of environmental effects on *health or socioeconomic conditions* can be taken into account in the EA with respect to Aboriginal peoples [s5.1c] or (when the designated project requires federal authority from an Act of Parliament other than the *Canadian Environmental Assessment Act*) with respect to other relevant groups [s5.2b]. It is also possible for health concerns to be addressed under [s19.1j], which refers to *any other matter* deemed by the responsible authority to be relevant to the EA.

(continued)

**Box 5.1** (continued)

*Step 5.* In order to conduct the EA in a manner that addresses the factors identified in Step 4, additional studies may be requested, most commonly by the responsible authority [s.23], but also by a review panel [s38-39], the Minister [s47.2] or the CEAA [s.106.2a]. This can involve the use or collection of any information, or undertaking of any studies (including those that the proponent is required to conduct) deemed necessary to conduct the EA, prepare the EA report or inform decisions regarding the EA report. The expert studies conducted at this step in the EA process tend to weigh heavily towards biophysical, ecosystem, and risk-based studies such as wildlife studies and ecological risk assessments (which evaluate the likelihood that adverse ecological effects may occur or are occurring as a result of exposure to one or more stressors, see US EPA 1992; Hope 2006). When health issues are identified in the scoping phase (Step 4), the most common type of assessments conducted in Step 5 are health and safety studies and human health risk assessments (HHRAs). Although HHRAs provide valuable information, their direct cause–effect, risk-based orientation is not designed to provide a full assessment of health impacts (see Briggs 2008), and tend to overlook the environmental, social, economic, or institutional factors that would be included in a "state of the art" HIA (Harris-Roxas et al. 2012, p. 43). Unless the expert studies conducted are well suited to a comprehensive assessment of health impacts, there is likely to be relatively insignificant application of the mandate to protect human health, as defined in the purpose of the *Canadian Environmental Assessment Act* [s4.2].

*Step 6.* The results of the expert studies that were conducted and any associated recommendations are forwarded to the responsible authority [s25.2 and 29.2], to the Minister [s52.3] and finally to the Governor in Council [s52.2] for a decision. The proposed project may be deemed to have no significant adverse environmental effects, or found to have adverse effects that can be justified [s31.1a(ii)] and [s52.4a]. In the latter case, conditions may be set for the project [s53] before the EA certificate that gives permission to proceed is granted. Rarely, a project may be considered to have unacceptable impacts, in which case it is rejected or returned to the proponent with a request for major revision.

The *Canadian Environmental Assessment Act* also includes a number of lesser or discretionary points that define when public participation can occur, with the potential to identify health concerns as factors to be considered within the scope of the assessment in Step 4 or as information relevant to the EA in Step 5 (e.g. [s24], [s28], [s43.1c], [s57] and [s58]). Although this opens the possibility for affected communities to have their health concerns included in the EA process, the weight given to this input by the responsible authority is not clearly defined. Since the *Canadian Environmental Assessment Act* was

(continued)

**Box 5.1** (continued)

revised in 2012, the implications for including human health considerations in the EA process have not yet undergone extensive legal challenges, so the full implications for the *Act*'s mandate to protect human health remain uncertain (see also Sect. 7.4). It is noteworthy, however, that even when a proposal successfully navigates the EA process, and an EA certificate is issued, there is no guarantee that projects will be implemented if potential adverse implications for environment, community and health have not been adequately addressed. Precedent-setting cases in the Supreme Court of Canada (1997, 2004, 2014) and recent court challenges (BC Nature 2014; Laanela 2014) are discussed below and important limits to EA processes are highlighted if they fail to account for the larger socioecological considerations that affect the health and well-being of affected communities (see also Chap. 2 and Vignette 6.6).

*Canadian Environmental Assessment Act* process is that health and a broader understanding of health impacts are clearly not the primary focus of the project or the assessment. Building from the description in Box 5.1 and the depiction in Fig. 5.1, it is evident that the legislation and relevant sections of the *Canadian Environmental Assessment Act* are open to interpretation and that, in many cases, health considerations are framed primarily in terms of the limited framework of an HHRA with an explicit focus on risks rather than impacts. HHRAs focus on classical, quantitative risk assessment, in which risk is calculated based on an assessment of sources, hazards, exposures, and known dose–response relationships, and therefore offer a very constrained perspective when compared against broader assessments of health impacts, such as that described by Briggs (2008, p. 3):

> Health impact assessment (HIA) provides an alternative paradigm. In contrast to risk assessment, this focuses on policies, or other interventions, rather than agents or events... It also recognizes that the environment is not just a hazard, but equally serves a beneficial role by providing natural capital... or ecological services... for example, through water security, improved nutrition or access to green space.

Canadian experience shows that if the community and health impacts of changes to the environment are not adequately addressed within reactive, legislatively driven EA processes, court challenges are likely to ensue. The 2014 approval of the Northern Gateway Pipeline project (following a Joint Review Panel process with the National Energy Board as responsible authority) is a notable example. Although extensive public submissions about interrelated environment, community, and health concerns were presented to the Joint Review Panel as part of the EA process, the Government of Canada accepted the panel's recommendation to approve the project (with 209 conditions, see Government of Canada 2014), and legal challenges promptly followed. BC First Nations filed appeals with the Federal Court seeking to overturn the panel recommendation (Laanela 2014), BC Nature issued notice of a lawsuit to challenge the federal Cabinet decision (BC Nature 2014) and

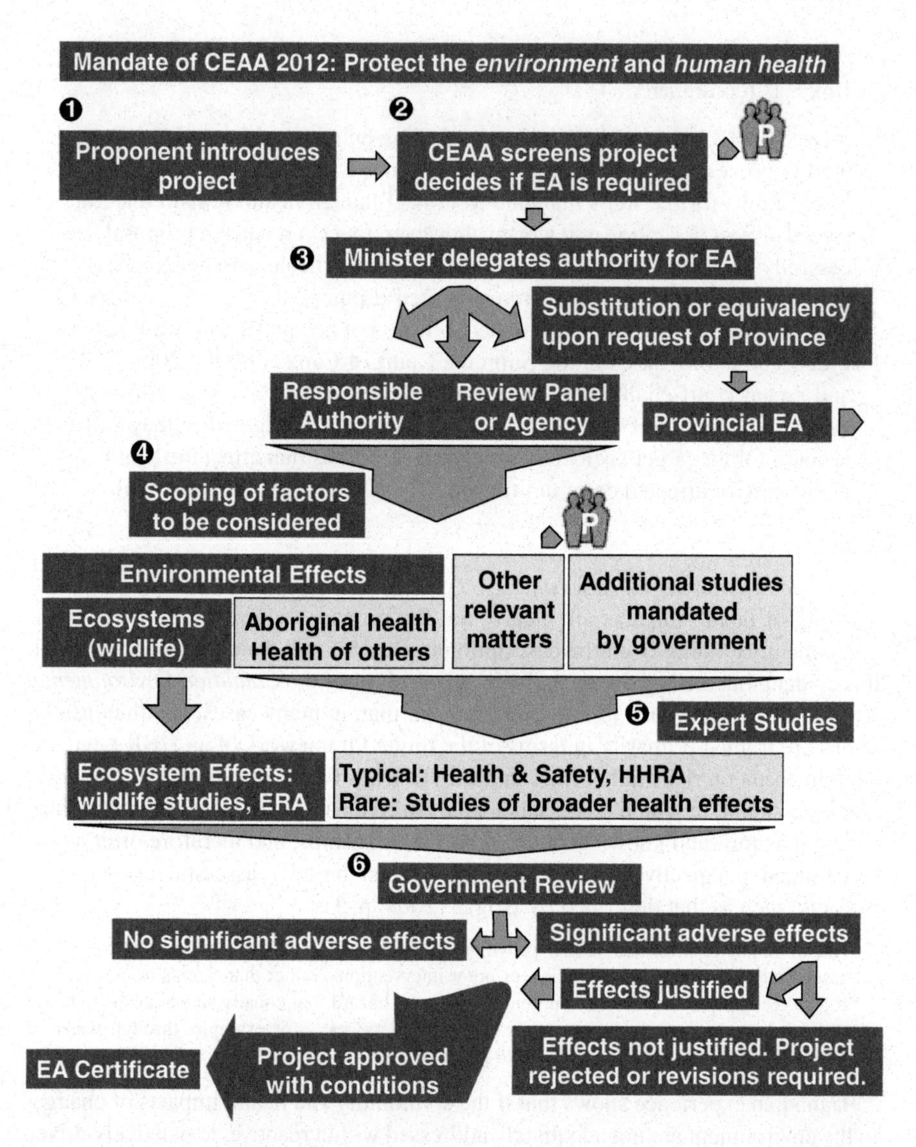


**Fig. 5.1** The *Canadian Environmental Assessment Act* (Government of Canada 2012) defines a process that has six major steps: (1) proposal and description of the project by a proponent; (2) screening by the CEAA; (3) delegation of authority for the review; (4) scoping to identify the factors to be considered; (5) expert studies and (6) decision on the project's fate. Within these steps, the *Canadian Environmental Assessment Act, 2012* identifies a series of discrete points where health considerations may be addressed (*yellow shading*), though these are not necessarily triggered in all environmental assessment processes. ERA, ecological risk assessment; HHRA, human health risk assessment; P, opportunity for *public* input where health concerns may be identified as factors to be considered in the EA. See Box 5.1 for details of these steps

the northern BC municipalities of Terrace, Prince Rupert, and Smithers voted to oppose the project. Precedent-setting court rulings in the Supreme Court of Canada related to Aboriginal rights and title have also highlighted significant implications for proponents moving forward with projects requiring an EA, especially in BC where few treaties with Aboriginal people's have been signed. Notable examples include: Delgamuukw v. British Columbia (Supreme Court of Canada 1997), which gave Aboriginal peoples the right to possess their ancestral lands; Haida Nation v. British Columbia (Supreme Court of Canada 2004), which required government (and, by extension, proponents) to engage in a meaningful process of consultation in good faith with First Nations; and Tsilhqot'in Nation v. British Columbia (Supreme Court of Canada 2014), which recognized First Nations title claims to lands they historically occupied, continually inhabited and exclusively use. These rulings are emerging as a force of evolution to the status quo EA process.

These Canadian court challenges, and others, underscore important limits to traditional project-specific and stressor-based EA processes in relation to cumulative environment, community, and health dynamics (see also Chaps. 6–8). Time spent in court challenging the inadequacies of current EA processes provides further impetus for more comprehensive approaches to assessing health impacts—especially given the expanding rate and scale of resource development projects, the associated social and ecological changes, and implications for the determinants of health. Such approaches are discussed in the next section with particular attention to the resource development context of northern Canada, although similar principles will be relevant elsewhere.

### 5.2.3 Emerging, Integrative Approaches to the Health Impacts of Resource Development

This section of the chapter highlights interrelated efforts that seek to include more integrative approaches to assessing health impacts, focusing especially on efforts that arise in direct response to resource development. These efforts emphasize the need to consider health impacts in ways that expand beyond formal or technical impact assessment processes, and the growing recognition of a cascade of direct and indirect impacts on individuals and communities, with health issues ranging from direct hazards and health service provision through to the social and environmental determinants of health in the short, medium, and long terms.

One type of effort to address the health impacts of resource development has focused on linking more integrative approaches within existing assessment processes, often by proposing a hybrid of the proactive and reactive approaches described in the two previous sections. Public concerns surrounding oil and gas development have provided an impetus and renewed attention to develop new approaches to including HIA as part of or in conjunction with EAs (Wernham 2007; Steingraber and Nolan 2012). Many of these examples have been associated with the resurgence in attention to more proactive approaches to HIA in Australia, Europe, and the USA (Bhatia 2007; Harris et al. 2012).

Yet in Canada, the lack of integration of health within environment impact assessments has been a central point of concern in several recent health-related assessments (see for example Gosselin et al. 2010; FBC 2012; OCMOH 2012). Specifically, a Royal Society of Canada report concluded in a recent assessment of the environmental and health impacts of Canada's oil sands industry that:

> The environmental impact assessment (EIA) process that is relied upon by decision-makers (i.e., panels for Alberta's Energy Resources Conservation Board, ERCB and in some cases Canadian Environmental Assessment Agency, CEAA) to make a determination whether proposals are in the public interest is seriously deficient in formal Health Impact Assessment (HIA) and quantitative socio-economic impact assessment (SEIA) as would be required for World Bank Projects, for example. Currently, human health impacts are assessed only by quantitative health risk assessment that is focused on predicting environmental contaminant exposures, while population health impacts (as outlined in the third bullet above), human health risk from technological disasters and occupational health are not addressed. Socio-economic impacts of developments are addressed only in a general, qualitative manner and these assessments would not satisfy the requirement of the World Bank for funding international development projects. Despite long-standing commitments to cumulative impacts assessment, there is little tangible progress evident in recent EIAs or current regulatory policy (Gosselin et al. 2010, p. 280)

In addition, despite growing calls for a more systematic inclusion of health considerations into EA, many oil and gas operations that will be operational for less than 150 days are very seldom reviewed in terms of health impacts, because they are considered temporary initiatives (Steingraber and Nolan 2012). The issue of time frames underscores the importance of understanding the different temporal orientations of health assessments. When considering different responses to the complexity of risk governance Briggs (2008) makes an important distinction between assessments that are *diagnostic* (i.e. that are required to determine the existence, magnitude and priority of a problem), *prognostic* (i.e. that evaluate and compare the implications of new or proposed policies and development projects) or *summative* (i.e. that evaluate the effectiveness of existing policies or development projects). The constrained framing of health considerations within the *Canadian Environmental Assessment Act* (see Box 5.1 and Fig. 5.1) exemplifies the limitations of a diagnostic approach (i.e. diagnosing problems associated with a single project) when this approach is not well linked to prognostic or summative processes. Indeed, the need for a shift towards more prognostic and summative orientations becomes especially important when considering cumulative impacts that may require a combination of considerations, across different time frames and priorities.

A 2007 review of cumulative environmental impacts and inequalities (Stephens et al. 2007) underscored the demand for new approaches to health impacts that extend beyond the limitations of most technical assessment processes. In addition to the limitations of the Canadian context (presented in Box 5.1 and Fig. 5.1), Stephens et al. (2007) focus on the limitations of existing methods to address health with EA processes. They highlight, for example, the limitations of methods that are commonly used within EIA frameworks, such as seeking expert opinions, developing checklists, performing spatial analyses, and modelling cross-sectional epidemiology, and especially note "problems with the use of linear models for understanding complex cumulative and multiple impacts, as the underlying relationships are frequently synergistic and iterative in nature" (Stephens et al. 2007, p. v). They also propose renewed attention to engaging with local community knowledge as well as expert knowledge, and emphasise the value of risk perception studies and of lay life-course mapping (asking individuals in affected communities to depict or describe the impacts of environmental change throughout different stages of their lives) in providing a more integrated appraisal.

Consistent with this emphasis, a second response to understand cumulative health impacts in the face of rapid and expanding resource development is associated with calls to move beyond formal EA or HIA processes in order to better reflect the full range of issues and implications for public health. A notable dilemma here is the lack of consistent or universally relevant frameworks for understanding the complex pathways by which resource development may influence health and well-being. Table 5.1 provides three contrasting, but complementary examples of the different languages and approaches that are emerging to frame and address the health impacts of resource development.

A 2012 example from northern BC (Table 5.1) reflects the issues identified by community members during the first phase of a diagnostic assessment commissioned by the BC Ministry of Health (FBC 2012). This first phase of the work focused on identifying health concerns related to oil and gas development in northeastern BC, and served as a basis to inform future phases of an HHRA (FBC 2012). The concerns identified by respondents exemplify a number of the challenges and disconnects described by Briggs (2008). Most notably, the health concerns identified in northeastern BC raised health concerns that extend well beyond the scope of issues that could be addressed by a quantitative HHRA (see also Box 5.1). Although HHRA was mandated as the assessment tool for the second phase of this process (Intrinsic 2015), this type of diagnostic assessment focuses almost entirely on issues categorised as environmental pathways of exposure (FBC 2012) and is poorly matched to respond to the range of other issues raised in phase 1 of this process. The Fraser Basin Council report (FBC 2012) underscores the question of what processes exist to respond to or address the other concerns identified in relation to oil and gas development in northeastern BC, which include personal health issues, related environmental issues, changes to the community, community service issues, and issues associated with gas operations, including exploration and drilling, transportation, and pipelines.

The need for more integrative approaches to understanding the range of health impacts in northern BC is reinforced by the contrasting perspectives in two reports released in the years following the FBC Phase 1 report in 2012. On the one hand, the health sector has called for renewed attention to the socioeconomic impacts of boom and bust cycles in northeastern BC and their implications for the determinants of health (Badenhorst et al. 2014; see also Chap. 4), thereby reinforcing the relevance of expanding attention to health equity considerations within health impact assessment processes (Povall et al. 2014). On the other hand, the 2015 release of the Phase 2 Human Health Risk Assessment (Intrinsic 2015) demonstrates the limitations of an assessment focused on the classic HHRA steps

**Table 5.1** Resource development and public health impacts in Australia and Canada: three examples that highlight the scope of issues, impacts and recommendations

| Context | Categories of issues, impacts and recommendations regarding public health and resource development |
| --- | --- |
| *Identifying Health Concerns Relating to Oil and Gas Development in Northeastern BC: Human Health Risk Assessment— Phase 1 Report*<br>BC, Canada (FBC 2012) | Categories of issues, based on respondents' concerns:<br>1. Personal health (physiological, psychological, cultural/spiritual)<br>2. Environmental pathways of exposure (air quality, water quality and quantity, food quality)<br>3. Related environmental issues (accidents, spills and explosions, increased traffic, noise and light pollution, impacts on the ecosystem)<br>4. Changes to the community (socioeconomic, demographic)<br>5. Community service issues (health care system, municipal and regional infrastructure, social and community services)<br>6. Oil and gas operational issues (exploration and drilling, processing and pipelines, transportation and traffic)<br>7. Institutional framework (monitoring and compliance, regulation and enforcement, communication, emergency response and planning, tracking and reporting) |
| *The Need to Measure and Manage the Cumulative Impacts of Resource Development on Public Health: an Australian Perspective*<br>Australia<br>(Kinnear et al. 2013) | Categories of cumulative public health impacts associated with the resource sector:<br>1. Health impacts for resource sector employees:<br>– Direct impacts (accidents, injuries)<br>– Indirect health impacts (increased risk of illness and disease)<br>2. Direct impacts on resource regions:<br>– Direct exposure pathways through air, soil and water pollution<br>3. Indirect impacts on resource regions:<br>– Flow-on impacts to partners, families, wider regional communities<br>4. Positive impacts and other pathways:<br>– A range of positive impacts ranging from lifestyle, well-being and socioeconomic benefits to increased awareness of safety and improvements to the health infrastructure and services |
| *Chief Medical Officer of Health's Recommendations Concerning Shale Gas Development in New Brunswick*<br>New Brunswick, Canada (OCMOH 2012) | Categories of recommendations for the protection of public health:<br>1. Protection of health and community well-being related to changes in the social environment (for example, recommendations for optimizing equitable distribution of risks and rewards and identifying a role for Public Health in community planning)<br>2. Protection of health related to changes in both the social and physical environments (for example, recommendations for developing a requirement to submit a health impact assessment (HIA) as part of the standard project registration process)<br>3. Protection of health related to changes in the physical environment (for example, recommendations for monitoring networks for ambient air and water quality)<br>4. Protection of future generations (for example, a plan to anticipate and mitigate *boomtown effects* and through strategic health impact assessments) |

of problem formulation, exposure assessment, toxicity assessment, and risk characterisation. Following the pattern of findings by the Royal Society of Canada report regarding the environmental and health impacts of Canada's oil sands industry (Gosselin et al. 2010, p. 280, quoted above), the HHRA approach adopted by the BC Ministry of Health in northeastern BC would not satisfy the formal HIA requirements required for World Bank development projects.

Each of the three examples in Table 5.1 reinforces the concerns raised by Briggs (2008) and others, namely that a mismatch can occur when health impacts are conceived of using narrowly defined causal pathways that directly influence individual health in a context where broader, long-term health implications should also be considered. Recent work from Australia (Kinnear et al. 2013) and New Brunswick, Canada (OCMOH 2012), both underscore the range of health-related concerns that cannot be addressed by the quantitative orientation of traditional risk assessments and highlight the need to also focus on indirect impacts associated with the broader determinants of population health. Both the New Brunswick and Australian examples (Table 5.1) highlight the challenge of identifying, measuring, and acting in ways that extend beyond the parameters of project-specific, reactive and risk-oriented approaches to assessing the implications for health. The recommendations proposed in New Brunswick (OCMOH 2012) identified a range of longer term considerations and challenges that span both the social and the physical environments, and demand collaboration and engagement that extend well beyond the health sector alone. The examples presented in Table 5.1 also reinforce the need for more integrative approaches to understanding health impacts, especially in ways that are better able to reflect the combined social and environmental determinants of health (Parkes et al. 2010; Hallstrom et al. 2015). The converging needs across all three examples highlight not only challenges relevant to understanding health impacts in isolation, but also the larger challenge of addressing the cumulative environmental, community and health impacts of resource development that are explored in the next section, and in other chapters of the book.

## 5.3 Cumulative Determinants of Health Impacts in Resource Development Contexts

The limitations and opportunities of existing approaches to assess health impacts become even more complex within the cumulative context of rapid social (social–demographic and socioeconomic) and environmental (biophysical and ecological) changes across northern Canadian landscapes. This section does not attempt to address the full scope of these issues, but does focus on the combined influence of cumulative impacts and determinants of health to draw attention to the concept of *cumulative determinants of health impacts* in rural, remote, and resource-dependent communities. Examples from northern BC highlight the regional context of direct and indirect health impacts, and the associated cascade of influences on the social and environmental determinants of health.

### 5.3.1 Direct and Indirect Health Impacts within a Regional Context of Cumulative Change

A central theme emerging from the examples discussed in the previous section is that discrete, project-driven health assessment processes are unlikely to adequately capture the range of health impacts of concern in contexts where multiple types of projects are occurring across the same landscapes and across a range of short, medium, and long time frames. A central challenge is the need to better reflect the direct and indirect pathways of health impacts on both individuals and the population of those who live within the affected areas. These dynamics are especially challenging in rural and remote areas, which are often the areas most likely to be hosting large-scale resource development.

Early work conducted by Frankish et al. (1996) found that HIAs in Canada were less likely to be undertaken in rural and remote areas, and were more likely to occur in suburban areas with higher and more densely situated populations. These issues point to challenges familiar in the field of rural and remote health, not least due to the fact that rural and remote areas are also likely to be characterised by low population density, limited data, and human resources that are already stretched across extensive landscapes and challenged by pre-existing social and structural determinants of health inequities (Smith et al. 2008; Bourke et al. 2010; Buykx et al. 2010). Issues related to measurability, quantification, and small populations compound and exacerbate the potential for health impacts to be invisible when dispersed across small communities within vast northern landscapes.

The examples provided in Table 5.1 also highlight important direct and indirect pathways that are especially relevant to rural and remote populations, ranging from direct individual impacts through to the social context of individuals who are working in the resource development sector, the sector-wide challenges of providing health services, as well as the need to pay attention to direct and indirect impacts that occur through longer term social and environmental determinants of health. In this context, the health sector is challenged to consider decisions that go well beyond individual projects, and that extend to questions of how to build a strategic, contextually informed, and long-term view. In this view, human risk and impact assessments are one of many considerations within a larger commitment to understanding cumulative health impacts within their areas of jurisdiction. The contextual detail provided in Box 5.2 provides a notable example of this view from the perspective of the Northern Health Authority, which serves northern BC in the context of rapidly expanding resource development.

The changing socioecological context described in Box 5.2 reiterates the challenge of understanding direct and indirect health impacts at temporal and spatial scales that go beyond those of individual projects. The scope of those impacts with potential relevance to the health sector is further developed by the integrative Vignettes described in Chap. 6, especially in relation to the role of the health benefits associated with: outdoor recreation and contact with nature (Vignette 6.5); the complexity of past, present, and future impacts on Aboriginal peoples (Vignette 6.6); and the importance of the health implications of the lived realities of local populations (Vignette 6.7).

**Box 5.2 Cumulative Health Impacts and the Northern Health Authority (Cathy Ulrich and Martha L. P. MacLeod)**

Northern Health (NH) is one of the five geographically based health authorities in BC that has the responsibility to provide health services, including acute care, home and community care, mental health and addiction care, and public and population health for a specific region of the province. The NH region, which covers the northern two-thirds of the province, is sparsely populated with about 300,000 people (or 6% of the provincial population).

The cumulative impacts of multiple natural resource development projects and the complex associated issues necessitates that NH take a partnership approach to understanding and acting on the challenges it faces, as reflected in its mission statement: "Through the efforts of our dedicated staff and physicians, in partnership with communities and organizations, we provide exceptional health services for Northerners" (Northern Health 2009, p. 2).

However, Northern Health has much to learn about the cumulative impacts of industrial development. There are three areas of work NH has undertaken to increase its knowledge and understanding of the challenges it faces. First, the Board of Directors leads a consultation process with industry every 2 years to increase awareness of the industrial development that is projected to occur, the associated health impacts and how industry will interface with the delivery of health services.

Second, NH recognises the need to understand the dynamics of industrial camps and their transient workforce. To this end, they have developed a snapshot of the number, size, and type of work camps and the associated transient populations. Alongside this work, NH has partnered with UNBC to review the historical patterns associated with resource and community development.

Third, NH has focused on learning from BC's experiences and those of other jurisdictions such as Alberta and Australia. There is a need for more documentation and research in this area to enable organisations such as NH to learn effectively from others.

Why is the topic of cumulative impacts of multiple natural resource development projects important for NH to understand? There are three ways in which NH is directly involved in and directly affected by such cumulative impacts.

First, NH is concerned about the health of the population. The population in northern BC experiences a poorer health status than their counterparts in the rest of the province. The opportunities and challenges that are inherent in economic development present a set of complex *wicked problems* (see Brown et al. 2010; Hallstrom et al. 2015). Many of these problems are beyond NH's capacity or mandate to resolve and require that multiple sectors come together to find solutions. For example, economic development brings employment and access to income, which are fundamental determinants of health and are therefore critical to health and well-being. At the same time, healthy environments (both the natural environment and the built environment) are critical to

(continued)

**Box 5.2** (continued)

health and well-being, as is the creation of resilient, vibrant, and healthy communities. The interplay and tensions among these and other determinants of health increase in complexity as industrial development increases. Northern Health is a partner in both leading and participating in addressing these tensions in ways that improve the health of the population. Such initiatives as the northern Road Health coalition, the Men's Health strategy, and the Partnering for Healthier Communities endeavour are examples of how NH has been involved (Northern Health 2010, 2012, 2014).

Second, NH experiences direct pressure on health services, its primary area of responsibility, as a result of both the construction and operational phases of industrial development. This is often experienced as increased requirements for after-hour care, pharmacy and diagnostic services, perinatal services as the birth rate increases due to the influx of a younger population, pre-hospital emergency services, trauma care to cope with workplace and recreational injuries, and mental health and substance-abuse services.

As communities grow and flourish with growing employment opportunities, NH finds the recruitment and retention of healthcare professionals to be easier. On the other hand, access to affordable housing and competition between employers for employees with the same skill sets create challenges for NH.

Finally, NH has direct responsibility for ensuring health and safety within a legislative and regulatory framework, such as protecting the quality of drinking and recreational water, and monitoring work camp infrastructure (including water, sewage and food facilities) under the *Drinking Water Protection Act* (Government of BC 2001) and the *Public Health Act*. (Government of BC 2008). Work in these areas increases as industrial development increases. There are several other regulatory areas that NH has an interest in, but no direct authority. These include air and watershed quality, noise, access to safe and affordable housing, access to day care spaces, and transportation infrastructure.

NH has a responsibility to provide information to industry during the EA process in relation to industrial development in northern BC, if and when health considerations are identified within the scope of factors to be taken into account. NH is then asked to analyse and respond from a health perspective to the project submissions from industry. These requests stretch NH's resources and technical capacity. In response to this demand, NH has recently created an Office of Health and Resource Development.

In summary, NH has experienced the cumulative impacts of multiple natural resource development projects through their effects on the health of the northern population, the health service delivery system, and the health and safety of the northern workforce. In response to these impacts, NH is working to understand and adapt to a dynamic and changing environment through partnered action, and a focus on learning, seeking to understand and participating in evaluation and research.

Although rare, precedents are emerging that focus on understanding the regional and population health impacts of resource development, taking into account both positive and negative influences on health (see Chap. 4, in relation to community dynamics). Wernham's (2007) HIA of oil and gas development on the Inupiat people of Alaska identified a combination of negative and positive health impacts. Consistent with the emphasis on positive and negative impacts in Box 5.2 (see especially Kinnear et al. 2013), Wernham's work identified negative health outcomes associated with changes in diet and lifestyle that would result from development of the land, an influx of temporary workers and exposure to pollutants, and also identified potential benefits related to increased economic well-being and the development of infrastructure and services (Wernham 2007). Literature is also emerging with a focus on the health impacts of mining operations on rural and remote populations, including attention to the specific dynamics of the mining resource sector in northern Canada (see, for example, Noble and Bronson 2005, 2006; Shandro et al. 2011). HIAs conducted in relation to international resource development projects continue to offer valuable points of reference to inform the Canadian context (see for example Utzinger et al. 2005; Snyder et al. 2012; Byambaa et al. 2014).

A growing body of research is highlighting the particular challenge of voicing integrated, culturally relevant concerns that connect health, environment, and food concerns within assessment processes and the challenges faced by Aboriginal peoples in responding to increasing resource extraction demands (Parlee et al. 2007; Tobin 2009; IHRC 2010; Place and Hanlon 2011; Shandro et al. 2014). Recurring themes include the extent of largely undocumented concerns that are scattered across northern Aboriginal communities, and the calls to develop HIA frameworks that are more relevant to Aboriginal peoples. Other insights about potential health impacts in Aboriginal communities in Canada can also be found in research with an emphasis on environmental contaminants (Lemire et al. 2015), and in unpublished reports funded by Health Canada's First Nations Environmental Contaminants Program (FNECP 2013).

Insights from rural, northern, and Aboriginal communities (Noble and Bronson 2005, 2006; Wernham 2007; Parlee et al. 2007; Shandro et al. 2014) highlight the importance of principles that were originally described in the mid- to late-1990s, whereby HIAs were envisioned as taking into account both direct and indirect impacts on the health of vulnerable populations, including the elderly, children, infants, pregnant women, workers, and Aboriginal peoples (Health Canada 1999). Attention to vulnerable populations brings to light some unique characteristics of the justice challenges faced by rural and remote communities. Much of the literature focused on HIA and environmental justice has tended to examine urban contexts, and specifically the environmental injustices experienced by the visible minority populations who are at greatest risk from living or working near urban and industrial projects (Krieg and Faber 2004; Stephens et al. 2007; Masuda et al. 2008; Wilson et al. 2008). This orientation is less applicable in the rural and remote context of northern BC, where injustices are not only experienced as a direct exposure to hazards by specific population groups, but are also associated with how the rural and remote communities affected by resource development are experiencing the decision-making and sociopolitical processes that are occurring around them. The examples in Box 5.3 provide vivid exemplars of these kinds of indirect impacts, with a specific focus on aging and senior populations.

**Box 5.3 Harnessing Planned, Sustainable Resource Development: Meeting the Needs of Northerners as They Age (Dawn Hemingway and Indrani Margolin)**

Northern BC has a wealth of natural resources—forests, coal, minerals, oil, gas, hydroelectric power, fishing, and agriculture. Our contribution to the provincial and national economy is enormous. Planned, safe, and sustainable development of these natural resources, including secondary and tertiary processing, has the potential to play a significant role in funding and maintaining social and physical infrastructures that are crucial to a better quality of life for northerners, especially those who are living on limited incomes or with the developing health challenges that come with age.

At the present time, however, much of the resource-generated wealth is funnelled away to enterprise owners, shareholders, and governments who are disconnected from the interests and needs of northern communities. The tendency is to *rip it and ship it*, with little or no processing. When resources are depleted, often all that is left is a hole in the ground, an abandoned mill or some other stranded asset. It is the communities of northern BC that are left to tackle the multiple challenges of this boom and bust economy. One growing concern is how this cyclical process can affect the health and well-being of senior citizens, especially in northern resource communities that are already underserviced (Hanlon and Halseth 2005).

Statistics reveal that the fastest-growing aging population in BC is located in the north (Northern Health 2013a, b). Research has shown that most of these older adults want to stay in their home communities (Northern Health 2013b). Thus, examining and addressing the impact of natural resource development on the health of these residents is necessary. Factors crucial to healthy aging in rural and urban northern BC include: (a) appropriate and affordable food, shelter and transportation; (b) available health care and social services; (c) accessible social connections and contacts; and (d) a positive environment that recognises and attends to specific age, gender, and cultural needs.

The cumulative impacts of the boom and bust approach to resource development can pose significant challenges to older adults, especially those in precarious situations regarding their income or health. Problems in boom times include shortages and high costs of appropriate housing, along with the high price of necessities such as food. In addition, older citizens are faced with social, community, and health services that are inadequate to support a rapidly growing population. Health care needs often cannot be met locally, and this requires relocation or expensive regional travel away from family and friends.

(continued)

**Box 5.3** (continued)

Bust times can be equally challenging. Although housing costs may decrease, other problems arise. For example, per capita government funding formulas can affect the availability of social infrastructure. Relying on population-based funding allocations in rural and remote areas with small and widely distributed populations inevitably results in more limited access to health care and other community-based resources in times of economic downturn. Exacerbating the situation, younger family members who leave to seek work elsewhere create additional challenges for older residents because the resulting population-based cuts in services are coupled with disruption of social support networks.

Despite these challenges, there are huge possibilities for successfully harnessing resource development to address the needs of seniors and their communities. Critical to this are decision-making processes that fully engage relevant community members and other stakeholders in discussion, and that enable a collective determination as to how natural resources should be extracted and processed to meet the needs of current and future generations. These stakeholders include Aboriginal peoples, resource sector workers, seniors' organisations, local communities, resource extraction and processing companies, and all levels of government. For example, in terms of access to natural resources, legislation could be enacted that favours companies and industries whose plans include further processing of resources in the north, thus adding breadth and depth to the local economy and alleviating some of the booms and busts. In addition, funding formulas could be developed that ensure that a percentage of resource dollars remain in the north to be invested in health, education, social services, and economic development. In order to achieve this vision, mechanisms are needed to ensure full community engagement. As one of the seniors' organisations in northern BC put it, "Nothing about us without us".

The focus on social and health inequities that are experienced by seniors and aging populations (Box 5.3) underscores recommendations by the New Brunswick Office of the Chief Medical Officer of Health regarding the need to protect health and community well-being in the context of a community's social and physical environments and also both within and across generations (see OCMOH 2012 and Table 5.1). Ensuring that explicit attention is paid to the direct and indirect pathways of health impacts across different population groups is consistent with calls for increased attention to the *causes of the causes* of health inequalities (Marmot 2005; CSDH 2008) and the recognition of the influence of ecosystems on human well-being through direct health impacts, ecosystem-mediated health impacts and indirect, deferred, and displaced health impacts (see Fig. SDM1 in Corvalan et al. 2005). The Health Synthesis section of the Millennium Ecosystem Assessment (Corvalan et al. 2005) expands these ideas further to emphasise the local, regional, and global changes that are influencing human well-being. In this wider context, the regional, dynamics in Boxes 5.3 and 5.4 can be understood in terms of direct and indirect drivers of change and their effects on both ecosystems and humans. This health synthesis also encourages the use of the concept of *ecosystem services* to understand how the effects and impacts on human well-being are related to the availability of basic materials for a good life, good social relations, security, and freedom of choice and action (Corvalan et al. 2005).

### 5.3.2 A Cascade of Impacts That Link Social and Environmental Determinants of Health

A recurring theme from the examples provided in previous sections is the need to understand the health impacts of resource development in relation to both the social *and* the environmental determinants of health. Clearly this need is not unique to understanding the health impacts of resource development, but is, instead, a reflection of a long-standing false dichotomy in public health discourse between the social and biophysical processes that influence health. In this dichotomy, when researchers, assessors, and decision-makers give attention to biophysical concerns, they tend to overlook social and cultural factors, and when they focus on social determinants of health, they tend to overlook the links between these determinants and related biophysical and ecological considerations (Cole et al. 1999; Parkes et al. 2003). Despite the progress that has been made in efforts to address this disconnect, the contemporary need to more fully understand the health impacts of resource development provides a renewed imperative to revisit and reinvigorate the notion of *reciprocal maintenance* that was proposed by the Ottawa Charter for Health Promotion in 1986 as a means to create supportive environments for health:

> Our societies are complex and interrelated. Health cannot be separated from other goals. The inextricable links between people and their environment constitute the basis for a socioecological approach to health. The overall guiding principle for the world, nations, regions and communities alike is the need to encourage reciprocal maintenance—to take care of each other, our communities and our natural environment. (WHO 1986, p. 3)

Although many would argue that the proposed socioecological approach to health has struggled to be realised in the more than 25 years since the Ottawa Charter (see, for example, Hancock 2011), the need to reconnect the social and ecological determinants of health has been revisited in a range of contexts in recent decades (Parkes et al. 2003; CPHA 2015), with the health impacts of resource development adding further momentum for this consideration. This thinking is consistent with long-standing calls to shift our attention *upstream* from the proximate determinants of health (McMichael 1999), with notable developments being articulated in the context of the determinants of Aboriginal health (Reading and Wien 2009). Greenwood and her colleagues offer an explicit integration of the social and biophysical determinants of health by embedding proximal, intermediate, and distal determinants of health within an integrated *web of being* that describes the interrelated factors that influence health (Greenwood and Place 2009; Greenwood and de Leeuw 2012, Fig. 1).

The interrelated social and environmental factors that determine health impacts become especially relevant when considered in the specific contexts of past, present, or proposed resource development projects or environmental changes. The examples provided in Box 5.4 provide a strong reminder of the spectrum of physical and psychological impacts that may occur within rapidly changing socioecological landscapes and which—due to ongoing impacts on the environment, communities, and health, especially within Aboriginal communities—may continue to have impacts across generations.

**Box 5.4 Mental Health and Well-Being Implications of Resource Development (Henry G. Harder)**

Why is it that when we examine the impact of an environmental change or development project, we acknowledge that it will have an impact on communities, but fail to give credence to the range and magnitude of those impacts? Most approaches to assessing these impacts don't seem to ask how communities are affected in terms other than environmental exposures or hazards. However, when a community is altered because of a pipeline, oil spill, or dam development, profound long-term change happens. The full dimensions of this phenomenon are rarely discussed and are certainly not featured in most discussions of health impacts.

Perhaps one of the most heinous resource development crimes in BC happened in the 1950s, when the Cheslatta Carrier Nation's territory was flooded by the building of the Kenney Dam to create the Nechako Reservoir. The federal and provincial governments of the day determined that this reservoir was needed in order to power the electric turbines that would provide electricity to the aluminum smelter in Kitimat, BC. Members of the Cheslatta Carrier Nation were forced off much of their land, including both their traditional hunting and fishing areas and their central village. The flooding that created the reservoir resulted in the destruction of their individual and community buildings as well as the graves of their ancestors. Some 60 years later, large water releases from the dam can still result in remnants surfacing from the desecrated graves. If one visits the Cheslatta Carrier Nation, the band office is full of pictures from that time. The memory is clearly still alive, as are the impacts of their experience. When one speaks to elders and other members of the community, many stories are recounted of persons who were unable to cope, and who fell into depression, anti-social behaviour, alcoholism or drug abuse.

The psychological impacts on communities and individuals are very real. Many will never recover. Some will not be able to cope. Not coping and not recovering have profound implications across the spectrum of mental illness, which can even include the potential for causing harm to themselves or others.

(continued)

**Box 5.4** (continued)

I believe that we need to make the public understand that resource extraction has a real and potentially deadly impact. What are the acceptable losses and the body count that industrial developers are willing to accept for their next project? Are we willing to accept their ability to profit from those impacts? The impacts described above are felt by all peoples. In Canada, however, such potential impacts are felt by Aboriginal peoples in disproportionate numbers. Resource extraction projects in Canada happen mainly in isolated areas where Aboriginal peoples are the dominant population. Current pipeline proposals in Canada, for example, are planned to cross traditional territories without due consideration of the impact on the persons who live in these areas. In fact, many communities have been affected more than once, and we must ask ourselves what the cumulative impacts of such experiences are. Albrecht et al. (2007, p. S95) have coined the term "solastalgia" to describe the phenomenon of loss, grief and distress caused by environmental change.

When society is deciding what the impacts of a resource development will be, the assessment must consider much more than direct, measurable, physical impacts, although these are also clearly critical. We must also ask what the psychological impacts will be both for communities and for individuals, and these impacts must weigh as heavily as the physical changes that are more easily measurable through traditional scientific techniques. Although making all impacts clearly visible to society would be a step in the right direction, this approach still allows society to decide that the cost is worth it and to make a decision based on narrow, selfish reasons. If we are to find decision pathways that are more suited to the long-term physical and psychological well-being of all citizens, we need to ask whether it is time to move away from the traditional, colonial, European–Christian model of resource extraction linked to the often-quoted Bible verse of Genesis 1:28:

> God blessed them and said to them, "Be fruitful and increase in number; fill the earth and subdue it. Rule over the fish in the sea and the birds in the sky and over every living creature that moves on the ground." (New International Version)

New pathways forward can be richly informed by embracing an Indigenous world view that focuses on stewardship and honouring the Earth and its resources. Embracing an Indigenous-oriented epistemology and axiology encourages us to move from a *subdue the Earth* stance towards a more cooperative, stewardship-oriented stance. Resource extraction could still occur, but would proceed with a genuine concern for the people living on the land, their past connection with the land and the future of that land, and the many species (including humans) whose lives depend on it. The Supreme Court of Canada's 2014 decision to give the Tsilhqot'in First Nation title to their traditional lands will have a profound effect on how future land use happens in Canada, and will

(continued)

**Box 5.4** (continued)

hopefully provide a clear example of Indigenous land stewardship in which family, individuals, and the sustainability of the Earth take precedence over fuelling continued colonisation in Canada and throughout the world (Supreme Court of Canada 2014).

There are important new opportunities to learn from such experiences, especially amidst the growing imperative of making decisions that value the physical and mental health of individuals and societies, as well as the ecosystems on which they depend.

The combined focus on mental health and well-being with an Indigenous perspective that is described in Box 5.4 reiterates the challenge of expanding our understanding of health impacts from a focus on direct (physical) health impacts on individuals that arise from specific projects to a broader, contextual understanding of direct and indirect health impacts that are influenced by interacting environmental and community factors. The metaphor of a cascade of impacts that has been proposed in this chapter to capture this idea builds on the well-known metaphor of *upstream* determinants of health to create an increased appreciation of how different dynamics of health impacts unfold over time and space in relation to direct and indirect impacts on the environment, the community, and individual health.

(Re)framing health impacts in relation to a cascade of impacts linked to the social and environmental determinants of health offers several points of reference to guide our understanding of and responses to the health impacts of resource development. Figure 5.2 illustrates the cascade, and depicts a range of intended metaphorical and conceptual features. The concepts of *connections* and *flows* are especially obvious—between upstream and downstream; between land and water; and among environmental, community, and health considerations. A cascade also conveys an important sense of directionality and pace associated with the flow from the top of a waterfall to its base. Although any lake or river is a dynamic system with ongoing processes of change, there is a notable, irreversible change—a tipping point—when a waterway becomes a cascade. Navigating the unpredictable (and likely tumultuous) journey downstream demands anticipation, planning, and preparation.

Working through the features of Fig. 5.2, the top of the cascade, where water tips over, offers a particular vantage point. From here, it is possible to observe and consider any *upstream* drivers of change in the landscape and waterways—including those being created by resource development—and how these combine, cumulatively, to feed into the cascade. Depending on the context and the pace of change, the view upstream could vary, from a calm lake or a wetland to a rapid, mountain-fed stream. From this same vantage point, it is also possible to look over and consider the (often unpredictable) downstream cascade that occurs once the flow of events from resource development begins to unfold. Visualising the downstream cascade encourages a recognition that impacts are likely to happen in a variety of

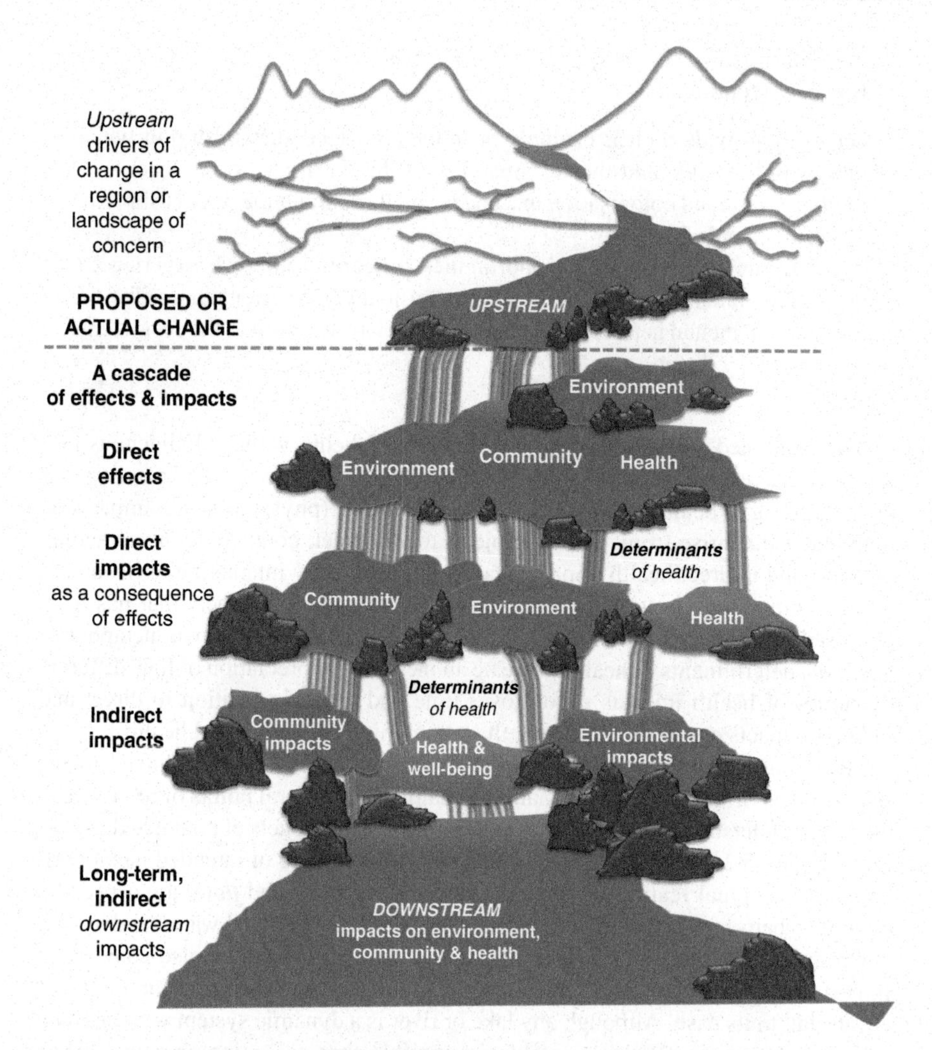


**Fig. 5.2** Resource development and the cascade of effects and impacts. Upstream drivers of change in a landscape or region of concern influence policies or projects that lead to a proposed or actual change. The integrative metaphor of a cascade depicts the downstream effects and impacts arising as a consequence of the proposed or actual change, and links community and environmental impacts with the social and environmental determinants of health

stages and with a range of possible pathways. Direct effects (on the environment) may divert the main flow to cause other direct and indirect impacts (on the wider community) that are associated with a range of social and environmental factors capable of influencing health and well-being.

The cascade image underscores the relationship between effects and impacts described in Chaps. 1 and 2, where an *effect* represents a change in the environment and an *impact* represents the consequences of such changes (Wärnbäck and Hilding-Rydevik 2009). The cascade metaphor depicts these direct and indirect relationships, whereby a direct consequence for the environment, community, or health may, in turn, lead to multiple indirect impacts farther downstream. These indirect impacts may be a product of time lags, threshold responses to environmental change or unintended consequences that subsequently influence the determinants of health. Specifically, the same drivers of change to upstream landscapes and waterways (whether through mining, forestry, agriculture, oil and gas, or other development projects) may have direct and interrelated impacts on both environments (Chap. 3) and communities (Chap. 4). These impacts can converge to influence the social and environmental determinants of health. Environmental determinants of health may be influenced through factors such as exposure to hazardous contaminants, degradation of soil or water quality, loss of access to landscapes or waterways or loss of habitat to conduct cultural practices and/or enjoy recreational opportunities; social determinants of health may be influenced by factors such as livelihoods, income, education, and access to social services and support (Parkes et al. 2010; see also Chap. 4).

Figure 5.2 depicts these interactions among the community, environment, and health by illustrating an array of potential downstream influences on health. The flows depicted in the figure also prompt a consideration of ways in which health impacts (including the provision or lack of healthcare) may also have indirect community and environmental consequences. This may be through the burdens communities face in caring for and supporting those in need of healthcare, or through the wastes and hazardous substances generated in hospitals and clinics, which are recognised as an environmental challenge the health sector must address.

The dynamic cascade depicted in Fig. 5.2 also prompts awareness that change can occur for a variety of reasons. Although the overall nature of a cascade can be surmised in an effort to navigate into the future, we must be attentive to variations in the quantity or pace of change (which could range, metaphorically, from droughts to floods), and also to obstructions or structural influences that may change the anticipated flows. Finally, at the base of the cascade, beyond the initial environmental, community, and health impacts, there is also a downstream flow that needs to be considered with an eye towards the medium- and long-term implications for the environment, communities, and related cumulative determinants of health impacts. Although not depicted explicitly in Fig. 5.2, consideration of these downstream dynamics also resonates with calls to consider socioecological implications for health, not only across time and across generations, but also between humans and other species (Waltner-Toews 2004; Wilcox et al. 2012; CPHA 2015).

## 5.4 Conclusions

Informed by the earlier literature and examples, this chapter explicitly reconnects cumulative impacts with the social and ecological determinants of health, and proposes a *cascade* of effects and impacts as an integrative metaphor for the *cumulative determinants of health impacts*. The focus on past, present, and future, and on upstream and downstream features (including both socioeconomic and ecological factors), provides a set of prompts to inform our understanding of the cumulative dynamics of environmental, community, and health impacts. The metaphor is also consistent with an understanding of health impacts that is informed by socioecological and (eco)systemic perspectives that recognise the interrelated health, ecological, and societal dynamics that are especially relevant in resource development contexts (see, for example, Webb et al. 2010; Poland and Dooris 2010; Charron 2012; Hallstrom et al. 2015).

Like any multifaceted metaphor, the cascade also poses complex challenges that should not be overlooked when seeking to address health impacts from this perspective. The direct and indirect pathways represented in Fig. 5.2 underscore long-standing methodological dilemmas associated with crossing scales, types of evidence, and knowledge cultures in ways that challenge standard scientific rules of evidence and that demand new approaches to integration and knowledge synthesis (Waltner-Toews 2004; see also Vignette 6.7). The interrelationships depicted and discussed throughout this chapter also demand critical reflection on the types of knowledge and voices that we value when seeking to understand health impacts, the tools we use to measure and integrate these understandings, and the processes by which we act on, reflect, and learn from what we know.

These integrative and cumulative challenges are not unique to health but, instead, are shared by environmental and community (societal and socioeconomic) dilemmas. In subsequent chapters, common opportunities, challenges, and innovations will be explored, building connections with a growing suite of research, practice, and policy efforts that are seeking to address the integration imperative across a range of health, ecological, and societal concerns. As a contribution to addressing this larger challenge, this chapter has highlighted the importance of health as an integral consideration in future impact assessments, and has also offered a variety of entry points for understanding and responding to the cumulative environmental, community, and health impacts of resource development.

## References

Albrecht, G.A., N. Higginbotham, P. Cashman, and K. Flint. 2007. Solastalgia: The distress caused by environmental change. *Australasian Psychiatry* 15: S95–S98.

Badenhorst, C., P. Mulroy, G. Thibault, and T. Healy. 2014. Reframing the conversation: Understanding socio-economic impact assessments within the cycles of boom and bust. *International Journal of Translation & Community Medicine* 2(3): 21–26.

BC Nature. 2014. BC nature to challenge cabinet decision approving northern gateway. Federation of BC naturalists. http://naturecanada.ca/news/bc-nature-to-challenge-cabinet-decision-to-approve-northern-gateway/. Accessed 9 Apr 2015.

Bhatia, R. 2007. Protecting health using an environmental impact assessment: A case study of San Francisco land use decisionmaking. *American Journal of Public Health* 97: 406–413.

Bhatia, R., and A. Wernham. 2008. Integrating human health into environmental impact assessment: An unrealized opportunity for environmental health and justice. *Environmental Health Perspectives* 116: 991–1000.

Bourke, L., J. Coffin, J. Fuller, and J. Taylor. 2010. Editorial: Rural health in Australia. *Rural Society* 20: 1–9.

Briggs, D.J. 2008. A framework for integrated environmental health impact assessment of systemic risks. *Environmental Health* 7: 61.

Brown, V.A., J.A. Harris, and J.Y. Russell. 2010. *Tackling wicked problems: Through the transdisciplinary imagination*. London: Earthscan.

Buykx, P., J. Humphreys, J. Wakerman, and D. Pashen. 2010. Systematic review of effective retention incentives for health workers in rural and remote areas: Towards evidence-based policy. *Australian Journal of Rural Health* 18: 102–109.

Byambaa, T., M. Wagler, and C.R. Janes. 2014. Bringing health impact assessment to the Mongolian resource sector: A story of successful diffusion. *Impact Assessment and Project Appraisal* 32: 241–245.

Charron, D.F. 2012. *Ecohealth research in practice: Innovative applications of an ecosystem approach to health*. New York: Springer.

Cole, D.C., J. Eyles, B.L. Gibson, and N. Ross. 1999. Links between humans and ecosystems: The implications of framing for health promotion strategies. *Health Promotion International* 14: 65–72.

Corburn, J. 2002. Environmental justice, local knowledge, and risk: The discourse of a community-based cumulative exposure assessment. *Environmental Management* 29: 451–466.

Corvalan, C., S. Hales, and A.J. McMichael. 2005. Ecosystems and human well-being: Health synthesis. In *Millennium ecosystem assessment*, ed. J. Sarukhán, A. Whyte, and MA Board of Review Editors. Geneva: World Health Organization.

CPHA (Canadian Public Health Association). 2015. *Global change and public health: Addressing the ecological determinants of health*. Ottawa: Canadian Public Health Association. www.cpha.ca/uploads/policy/edh-discussion_e.pdf. Accessed 30 May 2015.

CSDH (Commission on the Social Determinants of Health). 2008. *Closing the gap in a generation: Health equity through action on the social determinants of health; Final Report of the Commission on Social Determinants of Health*. Geneva: World Health Organization. http://whqlibdoc.who.int/publications/2008/9789241563703_eng.pdf. Accessed 9 Nov 2014.

Dannenberg, A.L., R. Bhatia, B.L. Cole, C. Dora, J.E. Fielding, K. Kraft, D. McClymont-Peace, J. Mindell, C. Onyekere, J.A. Roberts, C.L. Ross, C.D. Rutt, A. Scott-Samuel, and H.H. Tilson. 2006. Growing the field of health impact assessment in the United States: An agenda for research and practice. *American Journal of Public Health* 96: 262–270.

FBC (Fraser Basin Council). 2012. *Identifying health concerns relating to oil and gas development in Northeastern BC: Human health risk assessment—phase 1 report. Prepared for the BC ministry of health*. Victoria: BC Ministry of Health. http://www.health.gov.bc.ca/protect/oil-gas-assessment.html. Accessed 9 Nov 2014.

Fehr, R. 1999. Environmental Health impact assessment: Evaluation of a ten-step model. *Epidemiology* 10: 615–625.

FNECP (First Nations Environmental Contaminants Program). 2013. First Nations environmental contaminants program. http://www.environmentalcontaminants.ca. Accessed 9 Nov 2014.

Frankish, J., L.W. Green, P.A. Ratner, T. Chomik, and C. Larsen. 1996. *Health impact assessment as a tool for population health promotion and public policy. A report submitted to the health promotion development division of health Canada*. Vancouver: Institute of Health Promotion Research, University of British Columbia. http://catalogue.iugm.qc.ca/GEIDEFile/healthimpact.PDF?Archive=192469191064. Accessed 9 Nov 2014.

Gosselin, P., S.E. Hrudey, M.A. Naeth, A. Plourde, R. Therrien, G. Van Der Kraak, and Z. Xu. 2010. *The Royal Society of Canada expert panel: Environmental and health impacts of Canada's Oil sands industry*. Ottawa: Royal Society of Canada. https://rsc-src.ca/en/expert-panels/rsc-reports. Accessed 9 Nov 2014.

Government of BC. 2001. Drinking Water Protection Act, 2001 (S.B.C. 2001, C. 9). http://www.bclaws.ca/Recon/document/ID/freeside/00_01009_01. Accessed 1 Apr 2015.

———. 2008. Public Health Act, 2008 (S.B.C. 2008, C. 22) http://www.bclaws.ca/Recon/document/ID/freeside/00_08028_01. Accessed 12 Dec 2014.

Government of Canada. 1985. Fisheries Act, 1985 (R.S.C., 1985, c. F-14). http://laws-lois.justice.gc.ca/eng/acts/f-14/FullText.html. Accessed 14 Jan 2015.

———. 1994. Migratory Birds Convention Act, 1994 (S.C. 1994, C. 22). http://laws-lois.justice.gc.ca/eng/acts/M-7.01/FullText.html. Accessed 15 Jan 2015.

———. 2002. Species at Risk Act, 2002. (S.C. 2002, C. 29). http://laws-lois.justice.gc.ca/eng/acts/s-15.3/FullText.html. Accessed 10 Dec 2014.

———. 2012. Canadian Environmental Assessment Act, *2012* (S.C. 2012, C. 19, S. 52). http://laws-lois.justice.gc.ca/eng/acts/c-15.21/page-1.html. Accessed 12 Dec 2014.

———. 2014. *Government of Canada accepts recommendation to impose 209 conditions on northern gateway proposal*. Ottawa: Natural Resources Canada, News Releases. http://news.gc.ca/web/article-en.do?nid=858469. Accessed 10 Apr 2014.

Greenwood, M.L., and S.N. de Leeuw. 2012. Social determinants of health and the future well-being of Aboriginal children in Canada. *Paediatrics and Child Health* 17: 381–384.

Greenwood, M., and J. Place. 2009. Executive summary: The health of First Nations, Inuit and Metis children in Canada. In *Aboriginal children's health: Leaving no child behind.* Canadian Supplement to the State of the World's Children, ed. National Collaborating Centre for Aboriginal Health, 1–10. Toronto: UNICEF Canada.

Hallstrom, L., N. Guehlstorf, and M.W. Parkes. 2015. *Ecosystems, society and health: Pathways through diversity, convergence and integration*. Montreal: McGill-Queens University Press.

Hancock, T. 2011. It's the environment, stupid! Declining ecosystem health is THE threat to health in the 21st century. *Health Promotion International* 26: ii168–ii172.

Hanlon, N., and G. Halseth. 2005. The greying of resource communities in northern British Columbia: Implications for health care delivery in already-underserviced communities. *The Canadian Geographer* 49: 1–24.

Harris, P.J., L.A. Kemp, and P. Sainsbury. 2012. The essential elements of health impact assessment and healthy public policy: A qualitative study of practitioner perspectives. *BMJ Open* 2, e001245.

Harris-Roxas, B., and E. Harris. 2011. Differing forms, differing purposes: A typology of health impact assessment. *Environmental Impact Assessment Review* 31: 396–403.

Harris-Roxas, B., F. Viliani, A. Bond, B. Cave, M. Divall, P. Furu, P. Harris, M. Soeberg, A. Wernham, and M. Winkler. 2012. Health impact assessment: The state of the art. *Impact Assessment and Project Appraisal* 30: 43–52.

Health Canada. 1999. *A Canadian health impact assessment guide volume 1: The basics*. Ottawa: Health Canada. http://publications.gc.ca/collections/Collection/H46-2-99-235E-1.pdf. Accessed 5 Nov 2014.

———. 2004. *A Canadian health impact assessment guide volume 2: Approaches and decision-making*. Ottawa: Health Canada: Santé Canada. Archived June 24, 2013. http://publications.gc.ca/collections/Collection/H46-2-04-361E.pdf. Accessed 5 Nov 2014.

Hope, B.K. 2006. An examination of ecological risk assessment and management practices. *Environment International* 32: 983–995.

Huang, G., and J. London. 2012. Mapping cumulative environmental effects, social vulnerability, and health in the San Joaquin Valley, California. *American Journal of Public Health* 102: 830–832.

IHRC (International Human Rights Clinic). 2010. *Bearing the burden: The effects of mining on First Nations in British Columbia*. Cambridge: Harvard Law School. http://www.ceaa-acee.gc.ca/050/documents/p63928/92021E.pdf. Accessed 14 Nov 2014.

Intrinsic. 2015. *Detailed human health risk assessment. Phase 2 human health risk assessment of oil and gas activity in Northeastern British Columbia. Prepared for the BC ministry of health.* Victoria: BC Ministry of Health. http://www.health.gov.bc.ca/protect/oil-gas-assessment.html. Accessed 1 Apr 2015.

Kinnear, S., Z. Kabir, J. Mann, and L. Bricknell. 2013. The need to measure and manage the cumulative impacts of resource development on public health: An Australian perspective. In *Current topics in public health*, ed. A. Rodriguez-Morales, 125–144. Rijeka: InTech Publishers.

Krieg, E.J., and D.R. Faber. 2004. Not so black and white: Environmental justice and cumulative impact assessments. *Environmental Impact Assessment Review* 24: 667–694.

Laanela, M. 2014. Northern gateway pipeline: First Nations outline constitutional challenges: Grand Chief Stewart Phillip says at least 9 legal challenges have been launched. *CBC News*, July 14.http://www.cbc.ca/news/canada/british-columbia/northern-gateway-pipeline-first-nations-outline-constitutional-challenges-1.2706376. Accessed 8 Apr 2015.

Lee, J.H., N. Röbbel, and D.C. 2013. *Cross-country analysis of the institutionalization of health impact assessment*. Social Determinants of Health Discussion Paper Series 8 (Policy and Practice). Geneva: World Health Organization.

Lemire, M., M. Kwan, A.E. Laouan-Sidi, G. Muckle, C. Pirkle, P. Ayotte, and E. Dewailly. 2015. Local country food sources of methylmercury, selenium and omega-3 fatty acids in Nunavik, Northern Quebec. *The Science of the Total Environment* 509–510: 248–259.

Lewis, A.S., S.N. Sax, S.C. Wason, and S.L. Campleman. 2011. Non-chemical stressors and cumulative risk assessment: An overview of current initiatives and potential air pollutant interactions. *International Journal of Environmental Research and Public Health* 8: 2020–2073.

Marmot, M. 2005. Social determinants of health inequalities. *The Lancet* 365: 1099–1103.

Masuda, J.R., T. Zupancic, B. Poland, and D.C. Cole. 2008. Environmental health and vulnerable populations in Canada: Mapping an integrated equity-focused research agenda. *Canadian Geographer* 52: 427–450.

McCarthy, M., J.P. Biddulph, M. Utley, J. Ferguson, and S. Gallivan. 2002. A health impact assessment model for environmental changes attributable to development projects. *Journal of Epidemiology and Community Health* 56: 611–616.

McMichael, A.J. 1999. Prisoners of the proximate: Loosening the constraints on epidemiology in an age of change. *American Journal of Epidemiology* 149: 887–897.

Mendell, A. 2011. *Four types of impact assessment used in Canada.* Publication No: 1290. Quebec, Canada: National Collaborating Centre for Healthy Public Policy. http://www.ncchpp.ca/133/publications.ccnpps?id_article=581. Accessed 9 Jan 2015.

Mikkonen, J., and D. Raphael. 2010. *Social determinants of health: The Canadian facts.* Toronto: York University School of Health Policy and Management. http://www.thecanadianfacts.org. Accessed 9 Nov 2014.

Mindell, J.S., A. Boltong, and I. Forde. 2008. A review of health impact assessment frameworks. *Public Health* 122: 1177–1187.

Morgan, R.K. 2003. Health impact assessment: The wider context. *Bulletin of the World Health Organization* 81: 390.

———. 2011. Health and impact assessment: Are we seeing closer integration? *Environmental Impact Assessment Review* 31: 404–411.

Noble, B.F., and J.E. Bronson. 2005. Integrating human health into environmental impacts assessment: Case studies of Canada's northern mining resource sector. *Arctic* 58: 395–405.

———. 2006. Practitioner survey of the state of health integration in environmental assessment: The case of northern Canada. *Environmental Impact Assessment Review* 26: 410–424.

Northern Health. 2009. *Northern Health Strategic Plan* 2009–2015. http://www.northernhealth.ca/AboutUs/Mission,VisionStrategicPlan.aspx. Accessed 5 Nov 2014.

———. 2010. *Where are the Men? Chief medical officers report on the health and wellbeing of men and boys in Northern BC*. Prince George: Northern Health. http://men.northernhealth.ca/Resources.aspx. Accessed 15 Oct 2014.

———. 2013a. *Key issues in healthy aging: Strategies for health promotion. An integrated population health approach.* Prince George: Northern Health. Draft for Discussion Purposes. Version 1. September 2013. http://www.northernhealth.ca/Portals/0/About/PositionPapers/documents/HealthyAging_2013_09_V1_WEB.PDF. Accessed 9 Nov 2014.

———. 2013b. *Let's talk about healthy aging and seniors' wellness: Northern Health 2013 community consultation*. Prince George: Northern Health https://www.northernhealth.ca/Portals/0/

About/Community_Accountability/documents/March10-FINAL-report-on-2013-consultation-healthy-aging.pdf. Accessed 9 Nov 2014.

———. 2012. *Position on healthy communities. An integrated population health approach.* Prince George: Northern Health. http://www.northernhealth.ca/AboutUs/PositionStatements AddressingRiskFactors.aspx#532437-full-position-statements. Accessed 9 January 2015.

———. 2014. *IMAGINE collaboration: Collaborative solutions for health promotion: the road-health model.* Volume 3 of the IMAGINE Primer Series. MaryAnne Arcand & David Bowering (Guest Authors). Prince George: Northern Health. https://www.northernhealth.ca/Portals/0/Your_Health/Programs/Healthy%20Living%20And%20Communities/HealthyCommunities Toolkit/Imagine-Collaboration.pdf. Accessed 9 Jan 2015.

OCMOH (Office of the Chief Medical Health Officer). 2012. *Chief medical officer of health's recommendations concerning shale gas development in New Brunswick.* Fredericton: OCMOH. http://sustainabilityresearch.wp.rpi.edu/files/2013/04/NewBrunswickFrackingReport Sept2012.pdf. Accessed 14 Nov 2014.

Parkes, M., R. Panelli, and P. Weinstein. 2003. Converging paradigms for environmental health theory and practice. *Environmental Health Perspectives* 111: 669–675.

Parkes, M.W., K.E. Morrison, M.J. Bunch, L.K. Hallstrom, R.C. Neudoerffer, H.D. Venema, and D. Waltner-Toews. 2010. Towards integrated governance for water, health and social–ecological systems: The watershed governance prism. *Global Environmental Change* 20: 693–704.

Parlee, B., J. O'Neil, and Lutsel K'e Dene First Nation. 2007. The Dene way of life: Perspectives on health from Canada's north. *Journal of Canadian Studies* 41: 112–133.

Place, J., and N. Hanlon. 2011. Kill the lake? Kill the proposal: Accommodating First Nations' environmental values as a first step on the road to wellness. *GeoJournal* 76: 163–175.

Poland, B., and M. Dooris. 2010. A green and healthy future: The settings approach to building health, equity and sustainability. *Critical Public Health* 20: 281–298.

Povall, S.L., F.A. Haigh, D. Abrahams, and A. Scott-Samuel. 2014. Health equity impact assessment. *Health Promotion International* 29: 621–633.

Raphael, D. 2006. Social determinants of health: Present status, unanswered questions, and future directions. *International Journal of Health Services* 36: 651–677.

Reading, C., and F. Wien. 2009. *Health inequalities and social determinants of Aboriginal peoples' health.* Prince George: National Collaborating Centre for Aboriginal Health. http://www.nccah-ccnsa.ca/en/publications.aspx?sortcode=2.8.10&publication=46. Accessed 9 Nov 2014.

Schulz, A., and M. Northbridge. 2004. Social determinants of health: Implications for environmental health promotion. *Health Education and Behaviour* 31: 455–471.

Shandro, J.A., M.M. Veiga, J. Shoveller, M. Scoble, and M. Koehoorn. 2011. Perspectives on community health issues and the mining boom–bust cycle. *Resources Policy* 36: 178–186.

Shandro, J.A., L. Jokinen, K. Kerr, A.M. Sam, M. Scoble, and A. Ostry. 2014. *Ten Steps Ahead: Community Health and Safety in the Nak'al Bun/Stuart Lake Region During the Construction Phase of the Mount Milligan Mine.* Report by University of Victoria, Norman B. Keevil Institute of Mining Engineering, Monkey Forest Social Performance Consulting, Fort St James District, Nak'azdli Band Council. http://wildborderwatersheds.org/resources/category/mining. Accessed 15 Jan 2014.

Smith, K.B., J.S. Humphreys, and M.G.A. Wilson. 2008. Addressing the health disadvantage of rural populations: How does epidemiological evidence inform rural health policies and research? *Australian Journal of Rural Health* 16: 56–66.

Snyder, J., M. Wagler, O. Lkhagvasuren, L. Laing, C. Davison, and C. Janes. 2012. An equity tool for health impact assessments: Reflections from Mongolia. *Environmental Impact Assessment Review* 34: 83–91.

Steingraber, S., and K. Nolan. 2012. No compromise on NY fracking health impact assessment. *EcoWatch*, October 3. http://ecowatch.com/2012/no-compromise-on-ny-fracking-health-impact-assessment/. Accessed 9 Nov 2014.

Stephens, C., R. Willis, and G. Walker. 2007. *Addressing Environmental Inequalities: Cumulative Environmental Impacts.* Science report: SC020061/SR4. Bristol: Environment Agency. http://

www.staffs.ac.uk/schools/sciences/geography/links/IESR/downloads/cumulative%20 impacts%20full%20report.pdf. Accessed 9 Nov, 2014.

St-Pierre, L., and A. Mendell. 2012. Perspectives on health promotion from different areas of practice. In *Health promotion in Canada: Critical perspectives*, ed. I. Rootman, S. Dupéré, A. Pederson, and M. O'Neil, 117–137. Toronto: Canadian Scholar's Press.

Supreme Court of Canada. 1997. Delgamuukw v. British Columbia. 1997 3 SCR 1010. http://scc-csc.lexum.com/scc-csc/scc-csc/en/item/1569/index.do. Accessed 9 Apr 2015.

———. 2004. Haida Nation v. British Columbia (Minister of Forests), 2004 SCC 73 ed. 2004 SCC 73. http://scc-csc.lexum.com/scc-csc/scc-csc/en/item/2189/index.do. Accessed 9 Apr 2015.

———. 2014. Tsilhqot'in Nation v. British Columbia, 2014 SCC 44. http://scc-csc.lexum.com/scc-csc/scc-csc/en/item/14246/index.do. Accessed 9 Apr 2015.

Tobin, P. 2009. *Food security in the Takla Lake First Nation: Informing public health*. Saarbrücken: Lambert Academic Publishing.

US EPA (United States Environmental Protection Agency). 1992. *A Framework for Ecological Risk Assessment*. EPA/630/R-92/001. Washington DC: Risk Assessment Forum. http://www.epa.gov/osainter/raf/publications/pdfs/FRMWRK_ERA.PDF. Accessed 10 Apr 2015.

Utzinger, J., K. Wyss, D.D. Moto, N. Yémadji, M. Tanner, and B.H. Singer. 2005. Assessing health impacts of the Chad–Cameroon petroleum development and pipeline project: Challenges and a way forward. *Environmental Impact Assessment Review* 25: 63–93.

Waltner-Toews, D. 2004. *Ecosystem sustainability and health: A practical approach*. Cambridge: Cambridge University Press.

Wärnbäck, A., and T. Hilding-Rydevik. 2009. Cumulative effects in Swedish EIA practice—difficulties and obstacles. *Environmental Impact Assessment Review* 29: 107–115.

Webb, J., D. Mergler, M.W. Parkes, J. Saint-Charles, J. Spiegel, D. Waltner-Toews, A. Yassi, and R.F. Woollard. 2010. Tools for thoughtful action: The role of ecosystem approaches to health in enhancing public health. *Canadian Journal of Public Health* 101: 439–441.

Wernham, A. 2007. Inupiat health and proposed Alaskan oil development: Results of the first integrated health impact assessment/environmental impact statement for proposed oil development on Alaska's North slope. *EcoHealth* 4: 500–513.

———. 2011. Health impact assessments are needed in decision making about environmental and land-use policy. *Health Affairs* 30: 847–956.

White, H., P. O'Campo, R. Moineddin, and F. Matheson. 2013. Modeling the cumulative effects of social exposures on health: Moving beyond disease-specific models. *International Journal of Environmental Research and Public Health* 10: 1186–1201.

WHO (World Health Organization). 1986. *Ottawa charter for health promotion*. Geneva: World Health Organisation. http://www.who.int/healthpromotion/conferences/previous/ottawa/en. Accessed 9 Nov 2014.

———. 2014. Health impact assessment: Promoting health across all sectors of activity. http://www.who.int/hia/en/. Accessed 9 Nov 2014.

Wilcox, B.A., A.A. Aguirre, and P. Horwitz. 2012. Ecohealth: Connecting ecology, health, and sustainability. In *New directions in conservation medicine: Applied cases of ecological health*, ed. A.A. Aguirre, R.S. Ostfeld, and P. Daszak, 17–32. New York: Oxford University Press.

Wilson, S., M. Hutson, and M. Mujahid. 2008. How planning and zoning contribute to inequitable development, neighborhood health, and environmental injustice. *Environmental Justice* 1: 211–216.

Winkler, M.S., G.R. Krieger, M.J. Divall, B.H. Singer, and J. Utzinger. 2012. Health impact assessment of industrial development projects: A spatio-temporal visualization. *Geospatial Health* 6: 299.

Wismar, M., J. Blau, K. Ernst, and J. Figueras, eds. 2007. *The Effectiveness of Health impact assessment: Scope and limitations of supporting decision-making in Europe*. Copenhagen: World Health Organization on behalf of the European Observatory on Health Systems and Policies. http://www.euro.who.int/_data/assets/pdf_file/0003/98283/E90794.pdf. Accessed 9 Nov 2014.

# Part III
# Vignettes

# Chapter 6
# Exploring Cumulative Effects and Impacts Through Examples

**Michael P. Gillingham, Greg R. Halseth, Chris J. Johnson, and Margot W. Parkes**

*With contributions by* **Philip J. Burton, Alana J. Clason, Stephen J. Déry, Maya K. Gislason, Sybille Haeussler, Kathy J. Lewis, Nicole M. Lindsay, Kendra Mitchell-Foster, Phil M. Mullins, Bram F. Noble, Katherine L. Parker, Ian M. Picketts, Nobuya Suzuki, and Pamela A. Wright**

## 6.1 Introduction

In Part II we reviewed, from separate perspectives, many of the challenges posed by cumulative impacts relative to the environment, communities, and human health and well-being. Our goal with this book, however, is to develop a common framework for examining cumulative effects and impacts (see Box 1.1 in Chap. 1) inclusive of those environment, communities, and health and well-being perspectives. There are multiple drivers of change across landscapes, and the cumulative effects on a single place or species frequently lead to a cascade of impacts or consequences. Specific, on-the-ground examples can be very helpful to better understand the complexities, tensions, and interactions among cumulative impacts and foster more integrative approaches to addressing these. To that end, we present eight vignettes that help to illustrate recurrent messages about the failure of current thinking and approaches, and why we must develop a more inclusive perspective and ultimately integrative process for addressing cumulative impacts.

As noted in Chap. 3, determining the scale for examining cumulative impacts can sometimes be quite arbitrary. Watersheds, however, provide a convenient and appropriate scale of interest, and some of the most dramatic cumulative impacts are often associated with waterways and watersheds. The first vignette, *Exploring the Cumulative Impacts of Climate Change and Resource Development in the Nechako*

---

M.P. Gillingham (✉) • G.R. Halseth • C.J. Johnson • M.W. Parkes
The University of Northern British Columbia, Prince George, BC, Canada
e-mail: Michael.Gillingham@unbc.ca; Greg.Halseth@unbc.ca; Chris.Johnson@unbc.ca; Margot.Parkes@unbc.ca



M.P. Gillingham et al. (eds.), *The Integration Imperative*,
DOI 10.1007/978-3-319-22123-6_6

*Basin* (Picketts and Déry), examines the long history of changes in the Nechako Basin of northern BC—a watershed approaching the size of Switzerland—where in many ways little planning for cumulative impacts has been done, and where the values for considering cumulative impacts vary considerably.

One of the best ways to minimise impacts across watersheds is to identify ecological and human values *before* development takes place. Further, with those values identified, the opportunity to undertake scenario planning can be a key component of CEA (see Chap. 3). The second vignette, *Maintaining Wildlife and Wilderness in the Muskwa-Kechika Management Area* (Parker and Suzuki) describes a region in northern BC—an area of global significance—that was created as a result of a land-use planning process. In this area, provincial law requires that any industrial development in the area maintain wildlife and wilderness values (see Chap. 2).

To fully understand cumulative impacts, integration among potential ecological, economic, community, and health consequences must be more explicit. Many cumulative impacts originate with nonrenewable resource extraction. For renewable resources, as in the forest sector, long-term planning is normal, but plans cannot easily anticipate unexpected changes in the regional ecology or global economics. The third vignette, *Cumulative Effects and Impacts: Influence of the Resource-Based Business Model* (Lewis), uses examples of change in forest health to demonstrate that consideration of cumulative impacts must reach beyond just the consideration of economic activity.

The assessment of individual and of cumulative development impacts is almost by definition area based. In any one assessment, particular VCs/VECs may be of concern, but these assessments frequently overlook the cumulative impacts of activities on species and ecosystems that are not of economic importance. In some ways, the impacts on such VCs/VECs are intangibles. The fourth vignette, *Combating the Decline of Whitebark Pine Ecosystems across Central British Columbia* (Haeussler, Burton, and Clason), examines cumulative impacts for whitebark pine, a non-commercial tree species facing multiple interacting threats.

The ecological and social values of intact landscapes offer a range of benefits in terms of personal use and enjoyment, which are in turn affected by cumulative impacts. This is a complex reciprocal relationship whereby nature and nature-based activities have health and well-being benefits, but such activities—and the community and economic opportunities that bring people into these landscapes—can also, potentially, have impacts on the landscape. Vignette 5, *Valuing Outdoor Recreation in Living Landscapes as a means of Connecting Healthy Environments with Communities* (Mullins and Wright), explores the value of integrating nature into our daily lives through personal and societal choices and highlights facets of environmental values and consequences of landscape change that we frequently overlook.

Aboriginal peoples (see Chap. 1) have a complex relationship with the land involving both cultural and economic values. For Aboriginal peoples of northern BC, the cumulative impacts of European-induced landscape change began close to two centuries ago. Because the health, well-being, and identity of Aboriginal peoples are closely tied to the land (see Boxes 3.6 and 4.3), any consideration of cumulative impacts must consider how past, present, and future activities will influence not only the ecology, economy, and health of the people and their communities, but also culture and their relationship with place. Vignette 6, *Cumulative Environmental, Community and Health Impacts of Multiple Resource Development in Northern British Columbia: Focus on First Nations* (Lindsay), examines why the consideration of cumulative environmental, community, and health impacts from resource development must begin with an understanding of the significance of the land to Aboriginal communities.

Rapid landscape changes have a range of influences on the health and well-being of individuals and communities. In addition to direct causes of physical ill-health, there are complex and indirect impacts on mental health and well-being, and disruption to the long-term social and environmental determinants of health. The seventh vignette, *Lived Reality and Local Relevance: Complexity and Immediacy of Experienced Cumulative Long-term Impacts* (Mitchell-Foster and Gislason), suggests that broader vision and scope are needed to address the combined influence of communities, institutions, and natural systems on health and well-being.

Each of the first seven vignettes builds an argument, which will be expanded in Chaps. 7 and 8, that a dramatic shift in how we consider cumulative impacts is needed. Vignette 8, *Scoping Out Potentially Significant Impacts: Constraints of Current Regulatory-Based Cumulative Effects Assessment* (Noble), adds to this argument with illustrations of why current assessment approaches, which are focused on project approval, are ineffective.

## 6.2 Vignettes

***Vignette 1: Exploring the Cumulative Impacts of Climate Change and Resource Development in the Nechako Basin (Ian M. Picketts and Stephen J. Déry)***

The Nechako River originates in the coastal mountains in west-central BC and flows eastward for 440 km until it joins the Fraser River at Prince George, near the geographic centre of the province. It is an ecologically significant waterway, providing valuable habitat for a genetically distinct species of white sturgeon and also supporting multiple salmon runs. The Nechako River Basin (Fig. 6.1), which is comprised of the Nechako watershed and the Stuart-Takla watershed, is approximately 52,000 km$^2$ in size (as large as Switzerland), and is primarily made up of gently rolling coniferous-forested lands along with mountainous terrain in the basin's headwaters (BC Ministry of Forests 1998; Benke and Cushing 2005). The natural boundaries of a watershed offer an opportunity to clearly delineate and focus on a meaningful, albeit large, area that is easily understood, and relevant to natural and cultural values.

The Nechako Basin is the traditional territory of the Carrier people, who have been living in and travelling throughout the region for thousands of years. Traditional

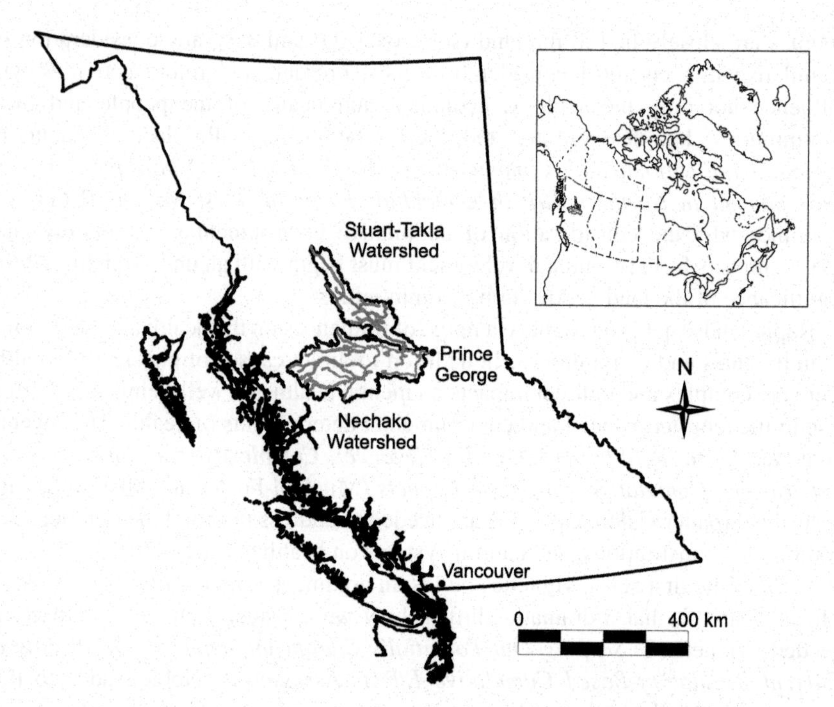


**Fig. 6.1** The Nechako Basin (major lakes and rivers shown in *blue*) drains both the Nechako and Stuart-Takla watersheds. Covering approximately 52,000 km$^2$, it encompasses primarily gently rolling coniferous-forested lands along with mountainous terrain in the basin's headwaters, and flows into the Fraser River at its confluence in Prince George, BC

diets include salmon, trout, moose, deer, and berries. In the 1700s, Europeans looking for furs began exploring the watershed, and in the 1900s Europeans began farming and logging activities in the basin. As population and resource development have increased in northern BC over the last century, the Nechako River, and the people who live within the watershed, have been subjected to substantial change. Major settlements such as Prince George, Vanderhoof, and Burns Lake have been established and grown within the river basin. These communities are primarily based on resource-related activities such as farming and forestry, and increasingly mining. Although many activities and events have affected the watershed, two major events have led to particularly profound changes to the Nechako River Basin, and have permanently affected the environmental and sociocultural health of the watershed. These are the following:

1. The construction of the Kenney Dam, completed in 1954 as part of the Kemano Development Project, created the 880 km$^2$ Nechako Reservoir, permanently dislocating hundreds of Cheslatta First Nations people, and redirecting much of the river's flow directly westward to power the Alcan aluminium smelter (see also Box 5.4). This significant watershed-level change has many lasting effects

related to water quantity and quality, channel morphology, hydrology, and fish species populations (Hartman 1996).

2. The current outbreak of the mountain pine beetle throughout much of BC has had profound impacts on the region's landscape and economy. This outbreak is expected to kill over half of the merchantable volume of pine trees in BC, and is greatly reducing the proportion of old-growth forests within the basin (Kurz et al. 2008). In an effort to find economic value in the dead trees, timber harvesting throughout the watershed has increased substantially in recent years (BC MoFLNRO 2012), but is projected to decrease substantially in the future as the logged and dead pine areas regenerate. Both the mountain pine beetle and the logging have major implications related to soil and nutrient cycles, wildlife populations, water availability, and erosion (Kurz et al. 2008).

Currently in BC there is a focus on encouraging resource development activities, particularly in the north. In the Nechako region there are major interests in additional mega-hydroelectric projects, oil and gas exploitation, petroleum and natural gas transport (i.e. pipelines), and several types of mining (including copper, gold, and molybdenum). Resource development is often promoted as a way to spur economic growth, provide jobs in northern communities, and help diversify the economy in regions that are beginning to experience a downturn in the forest industry (often related to the mountain pine beetle).

Climate change is now a current reality, and not just a future threat (IPCC 2013). People who live and work within a region must consider how past and future changes will affect the landscape and affect social, environmental, and economic viability. As a high-latitude inland area, climate change is occurring rapidly in north-central BC and is leading to changes related to water supply, natural ecosystems, and extreme weather events (Rodenhuis et al. 2009). One example of a climate-related change is the mountain pine beetle proliferation—consistently cold winter air temperatures, necessary for killing off the beetles, are no longer being experienced in central BC (Kurz et al. 2008). In light of the past and potential future changes and development in the Nechako Basin, people must be aware of the cumulative effects and resultant impacts of past, present, and future development activities, including climate change, in the watershed.

Climate change and resource development independently are complex sources of change and impact. Research at the watershed scale is meaningful, but can be challenging, as most information is collected at different spatial scales and can be difficult to apply. For example, forests are managed and forestry data are collected by Timber Supply Area in BC, and the Nechako River Basin contains portions of four different Timber Supply Areas. A significant amount of work, however, has already been done in the Nechako River Basin, particularly related to climate change impacts. This includes: climate change research that has focused on the resilience of forests and the forestry industry in the watershed (Williamson et al. 2008); identifying and addressing climate impacts in the City of Prince George (Picketts et al. 2013); examining climate change and water from both traditional and western perspectives with the Stellat'en First Nation (Sanderson et al. 2015); and conducting detailed climate modelling and hydrometeorological analysis of the Upper Fraser Basin (Shrestha et al. 2012). Based on the outputs of existing studies the major climate-related impacts in the Nechako River Basin have been found to relate to forest and aquatic ecosystem health, water supply, agriculture and food security, and community well-being.

Water supply is consistently identified as a priority concern in the region, and many forms of resource development (e.g. forestry and hydroelectric development) have significant impacts on water quality and quantity. As a system that is already disturbed, many stakeholders are highly concerned with ecosystem function in the watershed. Most forms of resource development (particularly those that disrupt river flow and that provide linear disturbances) affect terrestrial and aquatic ecosystems.

***Vignette 2: Maintaining Wildlife and Wilderness in the Muskwa-Kechika Management Area (Katherine L. Parker and Nobuya Suzuki)***

The Muskwa-Kechika Management Area (MKMA) in northeastern BC (Fig. 6.2) encompasses a region of scenic beauty, wilderness, wildlife, cultural values, and both renewable and nonrenewable resource potentials. Named after two major rivers (the Muskwa and the Kechika) that flow through it, the MKMA covers 6.4 million ha in the Rocky Mountains and is approximately the size of Ireland. The province of BC designated the MKMA as a distinct land base after adopting the recommendations of three community-based LRMP tables. The *Muskwa-Kechika Management Area Act* of 1998 (Government of BC 1998) thereby formalised the creation of the largest conservation system in North America (Shultis and Rutledge 2003). The MKMA has a conservation mandate to "maintain in perpetuity the wilderness quality, and the diversity and abundance of wildlife and the ecosystems on which it depends" (Government of BC 1998, under Preamble). It also has high potential for resource extraction and recreation, and therefore the challenge is to maintain environmental values with human activities on the landscape: "while allowing resource development and use in parts of the MKMA designated for those purposes, including recreation, hunting, trapping, timber harvesting, mineral exploration and mining, and oil and gas exploration and development" (Government of BC 1998, under Preamble). Furthermore, the long-term maintenance of wilderness characteristics, and wildlife and its habitat, is critical to the social and cultural well-being of nine First Nations that have territories within or adjacent to the MKMA (http://www.muskwa-kechika.com). To support the intent of achieving these goals, the MKMA includes parks and protected areas, as well as resource management areas where some resource development can occur as long as wildlife and wilderness values are maintained. It is guided by a unique management concept, trying to ensure a world-class standard for land use in a unique area.

To accommodate multiple shareholder involvement and perspectives, the provincial government established a Muskwa-Kechika Advisory Board of representatives

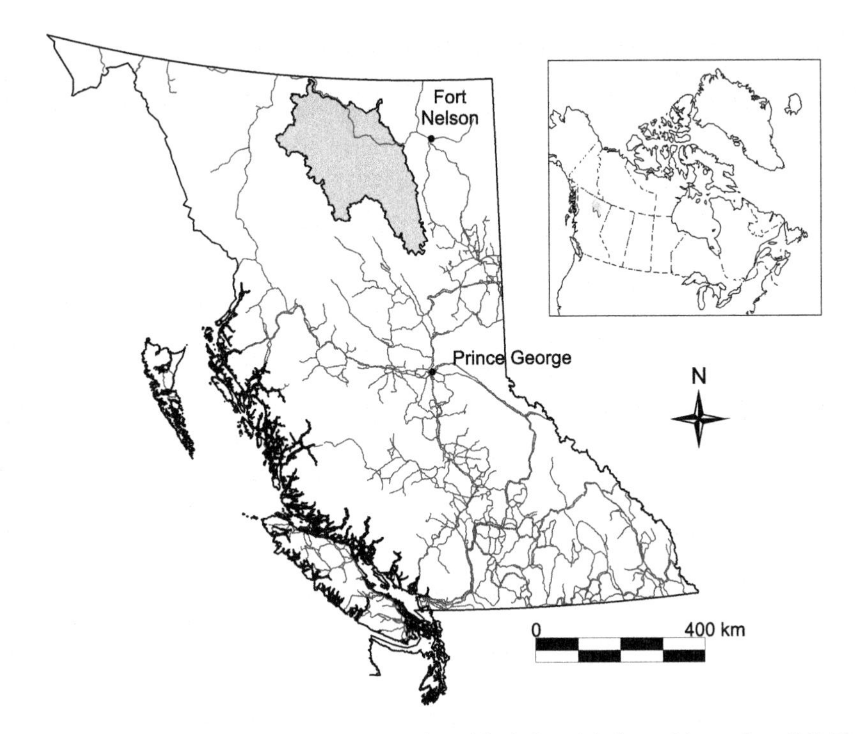


**Fig. 6.2** The Muskwa-Kechika Management Area (*shaded grey*) is located in northern BC. The area is comprised of a combination of provincial parks, protected areas, and special management areas (see text) and is a globally significant area of wilderness, wildlife, and cultures, to be maintained in perpetuity. Resource development and other human activities can occur in portions of the area provided that wildlife and wilderness values are maintained. Major roads are depicted throughout the province, but only a portion of the Alaska Highway intersects the Muskwa-Kechika Management Area

from a variety of interest groups to advise the government on natural resource management in the MKMA. To date, there has been almost no resource development or extraction in the area. Levels of human use and associated impacts, however, are likely to increase with improvement in the economic feasibility of mining, the anticipated expansion of oil and gas development from adjacent areas, the northward advance of beetle-killed forests, increasing demands for renewable resources (e.g. wind energy power), and heightened awareness of the area for tourism and recreation opportunities. To achieve the vision of the MKMA, it is imperative that anthropogenic effects are minimised and monitored closely, and that the resulting cumulative impacts do not substantively alter ecological integrity. As a frame of reference for managing cumulative impacts, a sustainable resource management strategy building on other planning activities for the MKMA was developed as the *Cumulative Effects*

*Assessment and Management Framework* (AXYS Environmental Consulting 2003; Salmo Consulting and Diversified Environmental Services 2003).

With the likelihood of increasing access into the MKMA, the biggest challenge will be to maintain its wildlife and wilderness values. The MKMA supports a unique diversity and abundance of wildlife, particularly ungulates and large carnivores. Intact large-mammal predator–prey systems at ecosystem scales exist because of the diversity in topography and biogeoclimatic zones, and almost no road access. It is important that the ecological pathways (i.e. linkages) among focal species and the importance of critical terrestrial and aquatic habitats towards maintaining biodiversity are understood well enough to minimise adverse impacts of increased human activities. Conservation tools directed towards this endeavour include a Conservation Area Design (Heinemeyer et al. 2004a, b), a Conservation Assessment that incorporates the implications of climate change (Yellowstone to Yukon Conservation Initiative 2012), and numerous applied research studies on components of the predator–prey landscape (e.g. Gustine et al. 2006; Milakovic and Parker 2011). Wilderness values in the MKMA include components of ecological integrity as well as human perceptions of naturalness. The culture and history of Aboriginal peoples are an integral part of this wilderness. Expectations of other users vary with mode of access and experiential valuation. This spectrum of local and visiting users must be recognised and acknowledged so that wilderness characteristics at landscape levels are assured for future generations.

Scenario planning offers the means to assess possible loss of wildlife and wilderness in areas where renewable and nonrenewable resource potentials are high. Spatially explicit projections of habitat value for wildlife, *habitat* values for recreationists and local users, and inputs from Aboriginal peoples can be considered in relation to projections of resource potential. Overlapping projections of conservation value and resource potential provide a heads-up for areas of possible conflict. Depending on the spatial and temporal extent of that conflict, some areas may need to be avoided whenever possible to meet the intent of the MKMA and in other areas activities might be conducted with relatively low impacts on ecological integrity. Strategic planning at this high level is useful in informing where conflicts might occur in the future for the MKMA and how landscape connectivity could be fragmented. To manage cumulative impacts, guidelines for land-use and habitat indicators and science-based thresholds for limits of acceptable changes will need to be defined.

It is a significant undertaking to operationalise the goals of maintaining wildlife and wilderness in perpetuity while still allowing resource development. Operational plans are probably of most value to industrial interests at a scale that specifically accommodates site-specific concerns. Soon after the MKMA was created, the provincial government initiated a series of pre-tenure plans to be completed prior to issuing oil and gas leases along the eastern edge of the MKMA. The pre-tenure plans, with input from biologists and local users, were designed to make industry aware of concerns related to ecological processes that could preclude *normal* operational policies and practices. The smaller scale covered by these plans—multiple watersheds rather than the entire MKMA—still embodies the premise of maintaining largely intact systems and requiring ecological restoration of the land base after development. The original oil and gas pre-tenure plans provided specific recommendations for spatially sensitive areas with time-sensitive access. More recent provincial regulations using a result-based approach eliminated the prescriptive nature of those plans in favour of placing the onus on industry to safeguard the conditions needed to retain wildlife systems and wilderness character. To ensure a coordinated approach to managing project-specific and cumulative impacts for the MKMA, however, resource management plans similar to the original oil and gas pre-tenure plans will need to be developed for each of the resource management zones, incorporating site-specific wildlife and wilderness information, guidance, and options for the planning, development, and restoration phases of any renewable or nonrenewable industry. These resource management plans must be accessible to resource users and regulators.

The size of the relatively isolated MKMA is its greatest asset. A rare opportunity still exists to truly integrate globally important ecological systems with land management activities and to sustain these systems as a whole even with economic development in parts of the MKMA. To succeed with this integration, next steps must define: (a) the extent of spatial scale that is needed to maintain ecological integrity across this landscape; (b) the time lag for environmental impacts that result from new activities; (c) how to assess the impact of a single proposed activity as well as the contribution of a single activity to cumulative change; and (d) the levels to which industrial and recreational access are adjusted to avoid adverse cumulative impacts of human activities on ecological integrity. Such an extensive area will also test the resolve to keep the vision of the MKMA. If the strength of this commitment is upheld and the plans and guidance to support decision-making are adhered to, the MKMA will indeed set a world-class standard for land use.

### *Vignette 3: Cumulative Effects and Impacts: Influence of the Resource-Based Business Model (Kathy J. Lewis)*

The potential ecological, economic, and social consequences that result from the combined individual effects of multiple resource-based activities over time have become of significant interest to those who manage and regulate industrial activities. Tools or systems for assessment and management of these cumulative impacts, however, are lacking. The BC Forest Practices Board in their 2011 (p. 4) report states simply that "we need to be able to assess the aggregate stresses acting on environmental values". Yet it is not so simple when one considers that these aggregate stresses are often driven by socioeconomic structures that are themselves cumulative effects resulting in a range of cumulative impacts. Any consideration or assessment of cumulative impacts must reach beyond the direct stresses to the environment brought about by industrial activities. Cumulative impact assessment must also examine other accumulated human activities, such as social acceptance of economic or business systems that provide the foundation for the resource-use activities, and the regulations and policies put in place to support these systems. The forest industry and emerging forest health problems as described in the following support this argument.

Healthy, productive forests are critically important for a continued supply of traditional economic products, and for the less tangible, but no less important, ecosystem services they provide. Ecosystem processes, such as disturbance and succession, help keep forests healthy and resilient as measured by variables such as habitat supply, biodiversity, and productivity. Insects, fungi, and fire are natural disturbance agents that have evolved with forest ecosystems and contribute significantly to forest diversity.

There are about 35,000 insect species in BC and over 10,000 known fungal species and some of these can and do kill trees. Despite the presence of these organisms, we still have forests due to a co-evolved balance between the host tree and an insect pest or fungal pathogen. Occasionally there is a shift in favour of the agent, but in functioning ecosystems, these are corrected by natural population controls, such as adverse weather conditions, or host vigour.

There is now considerable evidence that climate change is influencing the severity of damage caused by pathogens and insects. Circumstances have changed considerably since Waggoner (1962, p. 1100) defined a severe outbreak as a "rare removal by the weather of obstacles that ordinarily restrain the pathogen". Weather-related outbreaks are no longer rare and industrial activities and other human actions not directly related to land use are having significant impacts on forest health.

One example is the most recent outbreak of mountain pine beetle. Population size of this insect is regulated by weather (rate of beetle development), and by host condition (ability of trees to resist attack). This beetle is a natural part of forest ecosystems in western temperate forests, but in the past two decades these forests have experienced an outbreak of unprecedented proportions. Human activities resulting in climate change are one part of the story—since 1983 it has not been cold enough in winter to kill the larvae under the bark of infested trees. Human activities intended to protect the forest resource for lumber production are the other part of the story. Fire suppression since the mid-1900s created a predominance of older age classes of lodgepole pine and a lack of age class diversity on the landscape (Taylor and Carroll 2003).

A second example comes from a disease of lodgepole pine foliage, called red band needle blight. The population size of the fungal pathogen is regulated by weather as the spores are splash dispersed and need wet conditions for dispersal and infection, and by proximity of host trees (short dispersal distances). In northwestern BC, there has been a 20–30 % increase in mean summer precipitation in recent years compared to precipitation from 1960–1991, which has created much more favourable conditions for the fungus (Woods et al. 2005). This change alone would probably not have caused an outbreak of the disease, if lodgepole pine composition in the area was at the historic level of 9 % and forests were dominated by hemlock and subalpine fir (immune to the fungus). Species composition in much of this area, however, has swung heavily in favour of pine and spruce, which are preferred species due to their commercial value and fast growth.

In both cases, these organisms have caused outbreaks well beyond historical precedents, and in simple terms have resulted from changes in climate and a lack of diversity. An obvious solution for climate change adaptation is to increase forest diversity, but our ability to enhance diversity and, therefore, ecosystem resilience is significantly constrained by the dominant business model for forest resource use, and the policies and regulations that support the model.

We are economically and socially dependent on a business model that involves large-scale and extensive use of the forest to make large amounts of relatively low-value products (especially in the interior). Approximately 70 % of harvested log volume goes to medium and large lumber mills. These dimension lumber mills, and the pulp mills that utilise residual fibre, are limited by market demand and industrial systems designed to meet that demand, to two to three softwood species.

This very narrow business model and the perceived and real dependency of many rural communities on maintaining that model have led to an overemphasis of commercial species. BC has many different forest ecosystems with diverse natural forests that vary widely in species and age class composition; yet our reforested stands are much less diverse. There are two reasons for this. One is to provide future timber supply for dimensional lumber mills that only want to use three commercial species (spruce, pine, and Douglas fir). The second has to do with the requirement for licensees to restock harvested land, and until these sites are satisfactorily restocked, licensees are financially liable. Therefore, there is strong incentive to plant fast-growing, commercially desirable species. This has led to a significant narrowing of species diversity, which makes young forest stands more susceptible to damage by insects and fungi. Recent studies have shown that some reforested stands have fallen below minimum stocking levels due to insects and disease. Those stands will not meet expected timber yields for future timber supply (Woods and Coates 2013). Government and industry recognise that forests must be healthy and diverse in order to be able to adapt to changing climate and for continued productivity. Diversity is particularly important with maintaining resilience to biotic agents of disturbance because they tend to have a narrow host range.

The problem is that while diversity is critical for resilience, the current business model can only use three tree species. So forcing species diversity to enhance resilience will over time exacerbate existing timber supply problems. Therefore, the economic and social systems that are an integral part of our land-use practices are themselves causes of negative cumulative impacts on forest health and ultimately on the ability of forests to provide for human needs. Socioeconomic constructs, such as regulations and policies that support this narrow business model, constrain our ability to help forests adapt to changing climates, and these must be considered during cumulative impact assessments.

The solution requires serious investment in economic diversification and development of opportunities to use nontraditional species (e.g. hardwoods). Rather than trying to push ecological diversification for resilience to climate change, into a system where it does not fit, we need to create a market-based pull for diverse and higher value products from the forest. That will result in a business model that relies on forest diversity, thereby removing the negative cumulative impacts caused by regulations and policies that constrain diversity.

***Vignette 4: Combating the Decline of Whitebark Pine Ecosystems Across Central British Columbia (Sybille Haeussler, Philip J. Burton, and Alana J. Clason)***

Whitebark pine was declared endangered under Canada's *Species at Risk Act* (Government of Canada 2002) in 2012. This high-elevation tree is a keystone or foundation species in mountain ecosystems of western North America because its large, wingless seeds are exceptionally nutritious and because seed-caching by animals (primarily the Clark's nutcracker) initiates forest cover on harsh, exposed sites (Tomback et al. 2001).

Whitebark pine has been in precipitous decline over the past century due to the cumulative impacts and complex interactions of a warming climate, changes in forest succession and wildfires due to fire suppression and other human activities, repeated outbreaks of mountain pine beetle at ever higher elevations, and—most critically—the spread of an introduced Eurasian disease, the white pine blister rust. The loss of this VEC results in the unravelling of a vital montane-to-alpine food web that supports grizzly bears and many other wildlife species, and represents an important loss of ecosystem diversity in western mountain landscapes. BC's central interior, where whitebark pine reaches its northern geographic limit, is considered critical terrain for the future of this VEC because of the vast area of mountainous terrain to the north and west that may serve as suitable habitat under a warmer climate (McLane and Aitken 2012).

To understand the cumulative impacts contributing to whitebark pine decline we draw upon the concept developed by Manion (1981) and Mueller-Dombois (1988) who identified predisposing, triggering, and accelerating factors associated with biotic disturbances and forest decline (Fig. 6.3). Whitebark pine is entirely dependent on birds (Clark's nutcracker) for seed dispersal, is a poor competitor due to its slow growth rate and shade intolerance, and survives in harsh, mountainous terrain. These predisposing factors result in a patchy distribution with many isolated populations, inherently sensitive to rapid environmental change. Although a warming climate could open up new habitat beyond the current range limit, it may also enhance the growth of whitebark pine's subalpine competitors, pushing the pine out of current and new habitat. Formerly, mortality from fire and mountain pine beetle was offset by regeneration on the newly, open habitats. But the exotic blister rust is pushing this species over a threshold, perhaps into an extinction spiral (Fig. 6.3). Probable outcomes include species losses, reduced ecosystem carrying capacity, and ultimately a decline in the diversity of ecosystems across the subalpine landscape.

The decline of whitebark pine ecosystems represents a case study of cumulative environmental impacts that challenges the existing CEA framework and, more broadly, the governance of natural resources in Canada. Whitebark pine typically grows in high-elevation areas with few human inhabitants and little resource development. Human factors contributing to its decline are mostly diffuse and indirect

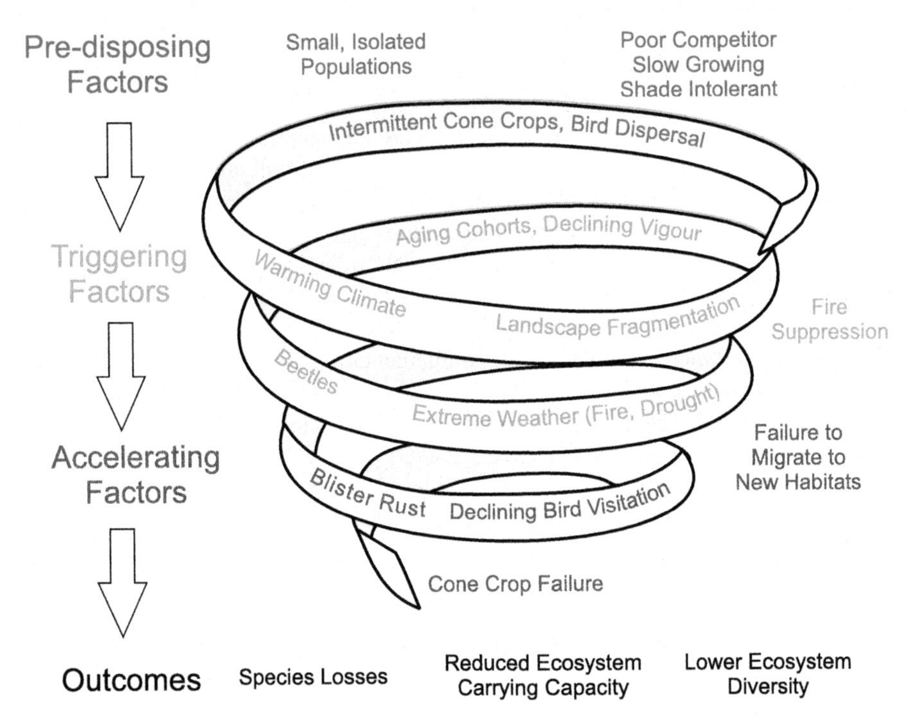


**Fig. 6.3** Cumulative environmental effects from natural and anthropogenic sources are causing rapid decline of whitebark pine populations and ecosystems. Feedbacks among predisposing (*in dark blue*), triggering (*light blue*), and accelerating factors (*in grey*; see text) may have precipitated an extinction spiral

(e.g. climate change, wildfire dynamics, and introduced disease) rather than direct and site specific (i.e. habitat destruction), making it difficult for governments to offload responsibility for environmental protection and restoration onto the private sector. On the other hand, development projects with a small footprint in whitebark pine habitat may have a disproportionate impact on the function, persistence, and renewal of this important ecosystem. For example, an energy corridor through a remote valley might result in stringent fire suppression in a landscape that previously could burn freely. Alternatively, a small development such as a mine or ski resort could reduce the number of seed-bearing whitebark pine trees in an isolated stand, decreasing Clark's nutcracker visitation rates, and thereby disrupt landscape connectivity and dispersal processes (Barringer et al. 2012). Industrial proponents and regulators will need to consider cumulative impacts over much broader spatial, temporal, and conceptual scales than they have until now.

Recovery efforts will require planting many millions of rust-resistant trees across hundreds of thousands of hectares of rugged mountain landscapes, as well as maintaining sufficient connected populations of seed-bearing trees to support Clark's nutcracker natural dispersal abilities. Natural and artificial selection may identify and promote rust-resistant genetic lines, but it will be a long road for populations to recover.

Because whitebark pine is not a commercial tree species and these ecosystems typically lie outside the timber-producing land base, they fall into a gap in jurisdiction between agencies with traditional responsibility for trees (BC Ministry of Forests, Lands, and Natural Resource Operations and the forest industry) and those accountable for wildlife, environmental protection, and parks who currently lack the capacity to address large-scale issues related to forest health, tree genetics, and reforestation. The federal government has no direct responsibility for the *Species at Risk Act* (Government of Canada 2002) implementation in central BC because there are no national parks in this part of the species' range. With whitebark pine ecosystems falling into these cracks in environmental jurisdiction and policy, a successful cumulative impact strategy for their recovery will require more collaborative arrangements involving governments, industrial and commercial enterprises, Aboriginal peoples, and environmental non-governmental organizations.

There has, however, been important recent progress in building such collaborative governance arrangements. Trees displaying signs of resistance to blister rust have been identified across central BC and seed collections for in situ and ex situ conservation have been made. The first ecosystem restoration and assisted migration trials involving genotypes adapted to a variety of climates and displaying blister-rust resistance have been established across a range of elevations (900–1800 m) from the Rocky Mountains, across the Nechako Plateau to the Coast Mountains. Support has come from regional environmental non-governmental organizations, universities, the provincial government, community forest corporations, Aboriginal peoples, mining companies, and charitable foundations. Forest licensees operating in central BC have set a goal of including 5 % whitebark pine in their tree-planting programmes within appropriate high-elevation ecosystems by 2017 (Mah and Astridge 2014). It remains to be seen, however, whether the recovery programme can scale up to the size needed to maintain functional whitebark pine ecosystems across central BC and to enable expansion northwards with the warming climate.

### *Vignette 5: Valuing Outdoor Recreation in Living Landscapes as a Means of Connecting Healthy Environments with Communities (Philip M. Mullins and Pamela A. Wright)*

Canadian landscapes include vast tracts of parks as well as public lands and waters, which are not designated as protected areas, but that nevertheless protect the environment and provide not only abundant resources for potential development, but also a broad array of ecosystem goods and services. Although forestry, mining, oil and gas, agriculture, and hydroelectricity are often the natural resources we might think of, outdoor recreation and tourism are also critical natural resources provided by these living landscapes. These landscapes are not just places to visit, trails for a hiker or mountain biker, or a favourite hunting or fishing spot. For people living in rural or northern regions, these landscapes define the *settings of everyday life* within which health is created and lived (WHO 1986). These *living* landscapes provide the context for the institutions with which we engage to study, work, and receive social services, and are also the places of work and play. The benefits of these living landscapes go far beyond the specific recreation setting or the tourism business, but provide myriad physical, mental, and social benefits to individuals and communities that support health and well-being.

Resource extraction and development can have far-reaching consequences for ecosystems and for community health and well-being. Related to such development, participation in outdoor recreation has been influenced by improved access, desire for healthy lifestyles, and declining environmental quality. Cumulative effect assessments should consider more fully the diverse landscape values that contribute to and shape recreation opportunities, the use of which produces outcomes related to individual well-being, community change and economic development, as well as the stewardship of environments, landscapes, and places. Recreation activities and benefits, however, are also in complex tension with broader social realities and other resource industries. Here we explore positive and negative impacts as well as the tensions, highlighting ways in which recreation-use values and benefits can be incorporated into the assessment of cumulative impacts.

Leisure, recreation, play, and physical activity have well-documented physical and psychological benefits for individuals and communities (Stephens 1988; Warburton et al. 2006). In northern and rural areas, these activities are often done outdoors. The health benefits of contact with nature have been identified in a growing body of work emerging from both the nature-environment sector, and the medical-health sector (e.g. Gies 2006; Kuo 2009). Furthermore, connecting with nature through outdoor recreation has significant community well-being and sustainability benefits including increased place attachment, levels of civic engagement, support for local heritage, as well as economic benefits comprising diversification, reduced health care needs, and employee retention (Canada Parks Council 2011; Government of BC 2011). Kruger (2006) also argues that outdoor recreation plays a significant role in place-making and community building for rural resource-dependant towns. Protection and investment in particularly attractive ecosystem services such as outdoor recreation opportunities can lead to amenity migration, in which people move to and invest in places with high recreation values for lifestyle reasons (Gartner and Lime 2000; Kruger 2006). Such migration may be attractive to rural communities as a stabilising influence given the economic fluctuations in the resource development sector. In the context of planning for future ecosystem services, Peterson et al. (2003) referred to amenity migration as a driver of business development and ecological change. Other impacts of amenity migration that could be considered in CEA include rapid inflation, degraded infrastructure, conflicting land-use values, ecosystem fragmentation, and pollution. As a demonstration of the complexity of the situation, these changes also introduce tensions within and across social classes and populations who differentially experience the costs and benefits of recreation and tourism resource development.

Facilitating community well-being by forging connections with local landscapes is particularly relevant for developing sustainable northern communities and industry sectors. To the extent that industry and government are focused on extraction of natural resources rather than stewarding natural landscapes and healthy ecosystems, they may jeopardise the future attractiveness and livability of rural and northern regions. Although little is currently known of the roles that outdoor recreation opportunities play, or could play, in building and maintaining healthy, strong, and unique communities that are attractive to local and immigrant populations over the long term, taking these health-nature connections into account is clearly an important consideration in future CEA.

The nature and type of participation in outdoor recreation and its diverse impacts on individuals and communities as well as the landscapes and environments they inhabit need to be understood, evaluated, and monitored as both a driver and recipient of cumulative impacts. The sociocultural, economic, and ecological consequences of developing outdoor recreation and nature-based tourism opportunities warrant explicit consideration in land-use decisions. Managing for sustainability and conservation also raises issues of social and environmental justice given that protected areas have also been noted to have significant social impacts including community displacement, loss of land uses, and visitor crowding (West et al. 2006). Meanwhile, recreation and tourism opportunities on public lands and their associated values and benefits may be altogether lost due to mining and forestry activity, for example. Nevertheless, road building for resource extraction can also induce growth in recreation use and tourism opportunities, with further effects (Hunt et al. 2009). This growth-inducing pathway is referred to as the Trojan horse effect, whereby initially low-volume and low-impact recreation or tourism development leads to further less manageable and more detrimental development (Butler 1990; see Chap. 2). It is also clear, however, that some forms of nature-based recreation and tourism often depend heavily on infrastructure developed initially for other resource industries.

To date, resource development plans (e.g. forest development/stewardship plans) or CEA have generally considered these forms of human use on the landscape with a relatively limited scope. In BC, site-specific tensions surrounding established recreation and tourism infrastructure (e.g. designated trails, campgrounds, and lodges) are typically examined, but consideration of the landscape or regional scale is limited and, when it is included, tends to focus on broadly established Visual Quality Objectives or Recreation Opportunity Spectrum polygons. Although these tools are an attempt to capture some human-use values, they are limited and incomplete ways of conceptualising the complex values related to human well-being within living landscapes. Additionally, Visual Quality Objectives or Recreation Opportunity Spectrum inventories are often incomplete and outdated, and consideration of these values is not required in many development projects, such as those for oil and gas

(BC Oil and Gas Commission 2014). A more comprehensive recreation and tourism inventory system that is updated regularly and planning and assessment processes that require attention to these values would enable more meaningful consideration of these values.

Similarly, the economic contributions of outdoor recreation and tourism values in a landscape often receive minimal if any consideration in resource development and CEA. Wilderness tourism in BC alone has reported direct tourist expenditures of approximately $1.5 billion dollars, representing 26,000 immediate full-time jobs and 40,000 jobs overall. Moreover, the broader BC tourism industry, representing in excess of $13.8 billion in revenues and 132,000 direct jobs, relies heavily on the values of *Super Natural British Columbia* provided by healthy and diverse landscapes (BC Wilderness Tourism Association 2013). Beyond tourism, valuation of the direct and indirect benefits of outdoor recreation and healthy lifestyles is significant, but largely unaccounted for in CEA. One recent examination of the economic contributions of officially designated BC Recreation Sites and Trails (Meyers Norris Penny 2011) reported economic impacts from operations generating $3.78 million in revenue and 300 full-time equivalent positions with an additional $18.5 million in revenues and 2400 full-time equivalent positions from user spending. This study also highlights other economic benefits from healthy lifestyles such as $4.4–$6.7 million annually in deferred health care costs from use of recreation sites and trails, diversification of local economies, and myriad other community and social benefits. Although limited in scope and scale, this report provides one example of how resource development and CEA could begin to more completely consider the human-use values of landscapes.

Cheng et al. (2003) described resource management as a form of place politics that natural and social scientists can inform with diverse and potentially otherwise-ignored values and meanings. Although not yet integrated into CEA, outdoor recreation and nature-based tourism research has shown such values, meanings, and outcomes and also developed relevant approaches to studying, measuring, and understanding them. For example, both quantitative and qualitative research have shown that outdoor recreation participation can increase participants' attachment to and identification with places and landscapes while encouraging environmental stewardship (Kruger 2006; Halpenny 2010). Measures of place attachment through outdoor recreation have become quite robust and are now becoming more complex (Hammitt et al. 2006; Oh et al. 2012; Budruk and Wilhelm Stanis 2013). Additionally, the study of recreation ecology typically examines site-level changes such as trail erosion, vegetation trampling, or wildlife disturbance in wilderness areas (Leung et al. 2008).

Recreation research could be adapted to better understand cumulative impacts and serve assessment processes in living landscapes. The roles of recreation and tourism in shaping place meanings and community priorities, desired limits to growth, as well as the impacts of other resource uses on citizen's attachment to and senses of place should be examined. Additionally, studies of recreation ecology could be expanded across diverse settings, to include transit zones as well as destinations for recreation and tourism, to the effects and impacts of recreation and tourism as well as on recreation and tourism, and finally to non-visible environmental changes such as mercury or lead levels in fish, for example. Operationally, BC's Adventure Tourism Policy demands rudimentary yearly reporting of use levels by businesses, which if enhanced could contribute to a cumulative impacts framework. Management for sustainability in living landscapes might doubly promote community well-being by first protecting ecosystem services upon which human and non-human communities depend, and secondly adding health and economic benefits through opportunities for ecologically compatible outdoor recreation and nature-based tourism (Millennium Ecosystem Assessment 2005).

### *Vignette 6: Cumulative Environmental, Community, and Health Impacts of Multiple Resource Developments in Northern British Columbia: Focus on First Nations (Nicole M. Lindsay)*

The health and well-being of Aboriginal peoples cannot be separated from the deep, complex, and enduring relationships they maintain with the land. Any consideration of the cumulative environmental, community, and health impacts of multiple resource development on Aboriginal peoples in northern BC must, therefore, begin with an understanding of the significance of the land as it relates not just to the livelihood and sustenance of many Aboriginal communities and individuals, but also to their cultural identities, languages, spiritualties, and experiences of health and wellness. For Aboriginal peoples, the land is much more than just *a collection of exploitable resources* (Booth and Skelton 2011); rather, it is linked to cultural identity, being, and wellness at a very deep level (Adelson 2000; Wilson 2003; Richmond et al. 2005; Richmond and Ross 2009; Parlee et al. 2007).

Equally fundamental in efforts towards understanding the cumulative impacts of resource development on Aboriginal peoples, however, is the recognition of how colonialism disrupts these integral relationships to land—and by extension disrupts and undermines Aboriginal connections to language, culture, identity, and well-being. Although many non-Aboriginal people would prefer to consider colonialism as solely historical or to deny its existence altogether,[1] Aboriginal peoples today—particularly those located on the frontiers of resource extraction—continue to experience colonialism and dispossession of land as a contemporary reality (Alfred and Corntassel 2005; Barker 2009; Hall 2012; Tuck and Yang 2012).

Colonialism, both historical and contemporary, is largely predicated upon securing access to natural resources. From the earliest waves of European traders and prospectors in search of furs and minerals through to the pioneering settlers, developers, and

[1] At a G20 Summit in September 2009 Canadian Prime Minister Stephen Harper stated that Canada has "no history of colonialism". Harper was widely criticised for this inaccurate characterisation of Canada's history—but the sentiment remains for many non-Aboriginal Canadians. For media coverage, see Ljunggren, D. "Every G20 nation wants to be Canada, insists PM" (Reuters, Sept. 25, 2009) and Hui, S. "Shawn Atleo criticizes Stephen Harper over *no history of colonialism* remark" (Strait.com, Oct. 2, 2009).

industrialists, colonial access to resources in Canada has been secured through a combination of formal treaty negotiations, widespread depopulation of Aboriginal peoples from introduced diseases, forcible removal of Aboriginal populations from their traditional territories to state-sanctioned reserves, suppression of traditional governance and cultural practices, and sustained assimilation efforts—including the removal of several generations of Aboriginal peoples' children into Residential Schools where they were forbidden to speak their native languages, and frequently subjected to mental, physical, and sexual abuse (Harris 2004). These colonial practices and processes have resulted in deep, lasting, and devastating effects on the social determinants of health for generations of Aboriginal communities in Canada, leading to drastically lower levels of health in virtually every indicator of wellness and disease (Adelson 2005; Gracey and King 2009; King et al. 2009; Reading and Wien 2009).

A wide body of research, much of it led by Indigenous researchers, highlights the crucial links between the health of the land and the health of Indigenous peoples around the world (LaDuke 1999; Wilson 2003; Windsor and McVey 2006; Richmond and Ross 2009; Berry et al. 2010; Richmond 2015). As Richmond (2015, p. 47) points out in a recent study of Anishinaabe Elders' stories of the relatedness of people, the land, and health, the "special, multifaceted relationship between Indigenous peoples and their local lands and traditional territories" is inseparable from their cultures, social relationships, and ways of living. The spiritual relatedness between Indigenous peoples and the land has been sustained through the transmission of Indigenous knowledge through countless generations, and is deeply rooted in Indigenous cultures. In Canada, colonialism and the resulting dispossession of and reduced access to traditional lands and territories have disrupted Aboriginal peoples' special relationships to the land, and can be seen as a key element of the cumulative impacts of colonial resource development.

According to Richmond (2015, p. 47), these changing relationships resulting from land dispossession and industrial development have had "cascading effects on the acquisition and practice of Indigenous knowledge, which has in turn affected the quality of social relationships—many of which are nurtured through time spent on the land—and compromised abilities to find and consume traditional foods". The impacts resonate through generations of Aboriginal people alienated from their lands, languages, and cultures.

Although it may be difficult for non-Aboriginal peoples to fully comprehend, the loss of access to traditional lands and resulting disruptions in cultural practices tied to the land, including the collection and use of *country* foods and medicines, has had deep ramifications that cannot be easily identified or qualified through Western scientific, medical, or historical methods. As noted by the authors of a study on Dene perspectives of health in northern Canada, these losses and changing relationships have both physical and non-physical effects. Parlee et al. (2007, p. 115) point out that "the destructive landscape changes associated with [large scale] resource development activities have tended to evoke strong sentiments of loss and alienation". Some community members interviewed by the researchers expressed concerns about the impacts of mining and hydroelectric development on the safety and health of consuming the traditional foods upon which they depend:

> It's going to get worse for the fish, caribou, and small animals that we trap when they develop other mines. The government is spoiling everything and the land will die … (J. Catholique, cited in Parlee et al. 2007, p. 124)

Taken together, the physical and non-physical cumulative impacts of industrial development run extraordinarily deep for the Dene people. The authors conclude by pointing out that:

> The capacity to achieve and maintain the Dene way of life—to pursue self-government, healing, and cultural preservation—in the face of other political, sociocultural, and environmental pressures is an ongoing challenge for Lutsel K'e Dene First Nation. The pressures associated with large-scale resource development activities are among the most pressing for many communities in the north. (Parlee et al. 2007, p. 130)

In short, Aboriginal communities in Canada have already been subjected to the cumulative impacts of resource-based colonialism for close to two centuries, although more recent rapid increases in the pace of large and mega-scale industrial development in the north are raising concerns to an alarmingly high level. A recent and devastating example of the industrial development risks borne by First Nations peoples in BC is the August 2014 tailings spill at Imperial Metals' Mt. Polley Mine in Secwepemc territory near Quesnel, in the Cariboo region of central BC. After the failure of a tailing pond dam, an estimated 10 million $m^3$ of mine waste water, 7.3 million $m^3$ of tailings (primarily finely ground rock), and 6.5 million $m^3$ of interstitial water (water suspended in the solid tailings) spilled into the Quesnel Lake watershed (Hoekstra 2014), a formerly pristine salmon-bearing system of rivers and lakes connected to the Fraser River. Following the disaster, a group of Secwepemc formed the Yuct Ne Senxiymetkwe camp near the entrance to the mine to monitor the impacts of the spill and remediation efforts on the part of the company. Although originally formed out of concern for the local impacts of the spill, the camp quickly built relationships of solidarity with other First Nations groups similarly resisting and raising awareness of the impacts of industrial development around the province. The Oct. 26, 2014, Yuct Ne Senxiymetkwe Facebook dispatch excerpted below clearly links together the broad and cumulative effects of colonial industrial development, as well as the rising tide of empowered First Nations resistance that aims to slow down, stop, and ultimately reverse these devastating effects:

> Today we are declaring victories. The Madii Lii encampment is a victory. Unist'ot'en Camp is a victory. Klabona Keepers are again and again victorious in their fights against industry and government. The Tsilhqot'in decision is a victory. We are building on the momentum and the power of all of these things. We are doing this in cities, we are doing this in villages, we are doing this on the street, we are doing it well and we are doing it consistently. It's time for better things. So build your Nations, stand strong on your Territories, it's time for new beginnings. We say colonization stops here. We say no more missing and murdered women. No more stolen children. No more concentration camp reservations. No more industry and industrialization. This is the end. This is the beginning. We are all fighting for something. (Yuct Ne Senxiymetkwe camp update, Oct. 26 2014)

As the powerful words above should make abundantly clear, First Nations peoples are not passive recipients of the cumulative impacts of industrial development. Rather, in order to adequately account for and address the broad range of cumulative effects and impacts of multiple resource development on First Nations in northern BC, their collective historical and contemporary experiences of colonialism and land dispossession, and their agency and resilience in resisting the abuses associated with this history and its attendant contemporary realities, must be understood. Additionally, and even more importantly, the aspirations, concerns, values, and ideas of First Nations people must not only be heard—they must take a central role in directing research and development policy priorities. First Nations communities in BC are diverse, and not all of them experience the effects of multiple resource development in the same way.

Researchers and proponents of resource development must also understand that past experience with these processes has been more negative than positive for many (if not most) First Nations peoples in northern BC, Aboriginal peoples in Canada, and Indigenous peoples globally. For many, ineffective and insufficient consultations, onerous and inadequate EA processes, broken promises, unanticipated harms, and unfulfilled expectations have undermined trust and good will (IHRC 2010; Booth and Skelton 2011; Place and Hanlon 2011). Given this history, we as researchers and advocates for Aboriginal Peoples health have a responsibility not just to account for and understand the cumulative impacts of resource development on Aboriginal peoples, but also to ensure that these efforts work towards *decolonizing* resource development by restoring good faith, nation-to-nation dialogue, and decision-making processes that prioritise the health and well-being of all of our relations for generations to come.

***Vignette 7: Lived Reality and Local Relevance: Complexity and Immediacy of Experienced Cumulative Long-Term Impacts (Kendra Mitchell-Foster and Maya K. Gislason)***

**Introduction**

Current Canadian federal mandates and economic growth plans based on resource extraction and exploitation for export globally are not designed with a prominent lens on health equity (FBC 2012; NRCAN 2013). Yet, it is widely accepted that intensive resource extraction processes, including the proliferation of new extractive technologies currently extending across geographic scales, affect human health at individual community and regional levels (Schmidt 2011; OCMOH 2012; Benusic 2013). The lived realities of people and communities, not limited to, but particularly at or near sites of extraction, are profoundly shaped by economic, political, ecological, cultural, spiritual, and social changes resulting from extractive industry activity (Sauve 2007; Parkes et al. 2011). Yet, formalised CEA processes have not been developed to reliably and meaningfully integrate the experiential and tacit knowledge held by communities that is supported by the evidence they gather from their daily lives (Brown et al. 2004; IPIECA 2005).

The myriad interacting factors and cumulative impacts of changes to landscapes, communities, and the cultures and psyches of communities embedded within these contexts of socioecological change should be measured, recorded, and shared in ways that reflect their complexity and help build an integrated understanding of health in the local context (Poland 2010; Sheppard et al. 2011). Individuals and communities who live close to intensive resource extraction sites have important insights to offer to such studies. For example, community wisdom can help articulate more nuanced insights into how health effects, as phenomena affecting biological (physical), psychological, and social aspects of human health (Adler 2009), are produced in the interplay between social-cultural, biological, and ecological determinants of health (Lee 2010; Parkes 2011).

Cumulative effect assessments, situated at the interface of governance, policy, and social–ecological systems, require a multidisciplinary and intersectoral approach in order to accommodate high degrees of complexity. One dimension of the complexity is that even the structuring of intersectoral spaces and collaborative relationships upon which much hope is pinned for producing new approaches to research also require an interrogation of their structures and functions in order to make explicit the retrenchment of dominant paradigms, historically entrenched knowledge biases, and inequitable power relations. The re-institutionalisation of inequitable conceptual, methodological, and practice-based orientations will likely occlude equitable community participation and meaningful integration of conventionally sidelined or subjugated knowledge (Foucault 2003; Fisher 2008). Equitable collaboration is a continuously evolving phenomenon in dialectic relationship with sociocultural and geopolitical processes. Sustainable development and community health are produced, to a degree, through these interactions and should be measured accordingly. Explicit inquiry, which deals conscientiously with interacting elements within social–ecological systems, is central to inclusive and exhaustive cumulative impact analysis and constitutes an important arena of research and practice (Parkes et al. 2010; Bunch et al. 2011).

Methodologies like community-based, participatory, and action-oriented research have sought to draw upon community knowledge to link environmental and community health concerns in nuanced ways (Ballard and Belsky 2010; Best and Holmes 2010). Yet, within biomedical and western scientific frameworks, evaluating and utilising diverse knowledge and integrating *non-technical* knowledge products and archives, like stories, songs, art, anecdote, collective imagination, vision, and cultural values, continue to perplex mainstream practice (Chandler and Lalonde 2004; Booth-Sweeny 2009; Kingsley et al. 2013). One way forward is to consider the importance of narrative and *micro-logics* as methodologically and scientifically rigorous contributions supporting more insightful and equitable integration of diverse ways of understanding the complex picture cumulative impact assessment is evolving to address. Knowing derived from embodied experience and local lived realities, for example, offers nuance and a *real-world* anchor to sophisticated scientific data-gathering processes; this helps to identify dynamics, sites, and processes of change that produce benefit or harm within specific places, spaces, and times (Struzik 2013).

## Micro-Logics, Equity, and Justice

In the case of provincially or federally contracted/sanctioned resource extraction and exploitation industries, economic macro-logics are often prime drivers of the scale and methods of resource development, distribution (i.e. export or domestic sale), and regulatory processes used to assess pre-, during, and post-impacts (O'Rourke and Connolly 2003). Macro-logics (i.e. strategic plans that are developed to function for a determined benefit at a large or *macro* political or social scale) by design do not consider benefits and harms produced within smaller socioecological and geopolitical systems (Barrington et al. 2012). An institutionally embedded system of inequity leaves unchallenged the practice of disassociating large-scale economic processes from the micro-logics or lived realities of communities where well-being is inextricably dependent upon ecosystem health (Greenwood and Place 2009; Krieger 2012). One of the key mechanisms of this erasure is the removal of non-technical reports and evidence of the immediacy around experienced harmful impacts in communities with urgent local relevance from formal risk assessment processes (First Nations Environmental Health Innovation Network 2008).

Ecological and social research methodologies hold that richness of information, and holistic understanding, around complex issues becomes progressively difficult to manage and integrate over increasing scales of inquiry. One current running through this challenge is that cultural and social ways of knowing, particularly for rural, remote, northern, and Aboriginal communities, are embedded in landscapes and as the geographical scales across which an issue is proliferating so too is the depth, breadth, and richness of social, cultural, and lived knowledge pertaining to the issues. As such, developing robust local capacities for complex and long-term measurement of impacts, for critical examination of locally relevant knowledge and urgent community issues, and for meaningfully integrating locally cultivated evidence into meso- and macro-level processes is imperative (Barrington et al. 2012). A closer look into how information and knowledge are gathered, used, and mobilised by and for cumulative impact assessments is required, especially with a strong and intentional emphasis on addressing these gaps in research, practice, and policy. Further, the dearth of inquiry in examining the relationships between the intertwined micro-, meso-, and macro-logics that deal with psychological, spiritual, social, cultural, and *familial* cumulative impacts of these development projects needs to be expressed within these measurement matrices (Boyle and Dowlatabadi 2011). Important questions to ask include the following: "What are the micro-logics relevant to oil and gas exploration and extractive policy in northern BC?" and "How do the particular sociocultural and geo-political narratives give rise to increased complexity when examining these social–ecological systems?" as they are *game changers* for many communities in BC at present. Micro-logics, it warrants noting, also include the adaptive processes of communities to changes in pressures and exposures that determine health, well-being, and quality of life (Boyle and Dowlatabadi 2011).

Given that these are deeply rooted and far-reaching problems, the strategies used to mitigate problems may themselves give rise to new and unexpected impacts and produce a novel set of factors that also need to be identified and measured (Brown et al. 2010). For example, environmental degradation in the form of the contamination of watersheds may lead to a depletion of fish stocks and wild game, and decreased agricultural capacity, among other effects. Communities may adapt to these pressures by abandoning traditional ways of living (i.e. fisheries, livestock farming, hunting) with devastating consequences for erosion of cultural identity, family structure, and community social fabric—all crucial dimensions to human health and well-being when understood to be a fully encompassing biopsychosocial phenomenon (Stephen 2013).

Successive webs of impact are difficult to articulate, and even to understand, especially for individuals and groups whose ways of living, working, and knowing do not intersect with the experience of those living with multiple and cumulative impacts on health and well-being (Boyle and Dowlatabadi 2011). Barring an ideal scenario of a paradigmatic shift towards macro-logic determination through equitable integration of multiple micro-logics linked by natural social–ecological systems (Keskinen et al. 2012), practical steps must be taken to develop methodologies and tools that integrate focuses on health equity, social justice, environmental justice, and equitable community collaboration for cumulative health and environmental impact assessments (Barzyk et al. 2010; Sexton 2012).

Critical social theory and activism ethnography present themselves as candidates for chronicling the evolving nature of relationships between groups of people, institutions, public and environmental health, and governance. Community-based, participatory, and equity-framed analyses of priorities, power-sharing, and power differentials between and among groups, institutions, and logics may provide a much-needed locally relevant foundation to support policy and legal analyses from multiple sectors, scales, and socioecological contexts. A broader scope and vision are needed for current cumulative impact analyses to meet the combined needs of communities, institutions, and natural systems. Relatedly, more inclusive and collaborative orientations, approaches, and methodologies are needed to adequately address the systematic exclusion of nuanced experiential, tacit, and community-based knowledge from intensive resource extraction processes.

### *Vignette 8: Scoping Out Potentially Significant Impacts: Constraints of Current Regulatory-Based Cumulative Effects Assessment (Bram F. Noble)*

The primary, legislated instrument assessing and managing the cumulative effects of resource development is project-based EA. Federally, CEA is a requirement under the *Canadian Environmental Assessment Act, 2012* (Government of Canada

2012). Section 19(1)(a) of the Act requires that the EA of a designated project consider, amongst other factors, any cumulative environmental effects that are likely to result from the project in combination with other physical activities that have been or will be carried out. The assessment of cumulative effects is also variably considered in each of Canada's provinces and territories for project-specific development proposals—either through provincial EA legislation, EA regulations, or on a case-by-case basis as part of project-specific guidelines or terms of reference.

The basic challenge to current CEA provisions in Canada, however, is that thinking cumulatively does not emerge naturally from a project-based perspective (Parkins 2011). Project-based EA in Canada has been widely criticised for its approach to cumulative effects due, in part, to: (a) many projects being considered too small or insignificant to trigger legislated or regulatory requirements for assessment, and, therefore, no assessment is undertaken and cumulative effects go unchecked (Nielsen et al. 2012); (b) the narrow spatial and temporal scope of project-based EAs, which do not take into account the full extent of the impacts of multiple stressors on biophysical and socioeconomic environments over space and time (Seitz et al. 2011); (c) the inability for individual proponents to manage effects beyond the scope of their own projects, because they have neither the mandate nor the authority to do so (Baxter et al. 2001); and (d) the ultimate focus of EA is making sure that the impacts of a project are acceptably small rather than understanding the total effects of all stressors, project and non-project, on any single environmental component (Duinker and Greig 2006).

In the sections that follow two examples of recent regulatory-based CEA are provided that illustrate the constraints of the project-based model. Although focused on hydroelectric development in northern Manitoba and natural gas development in northern BC, both under provincial EA systems, the two cases are typical examples of the challenges to project-based CEA across Canada (see Duinker and Greig 2006; Parkins 2011; Seitz et al. 2011; Dubé and Wilson 2013).

#### Example 1: Bipole III Hydroelectric Transmission Line Project, Manitoba

In 2011, Manitoba Hydro, a Crown Corporation of the Government of Manitoba, filed an environmental impact statement with the Manitoba Clean Environment Commission (MCEC) for the construction and operation of the Bipole III high-voltage direct current transmission project. The proposed project involves the construction of an approximately 500 kV, 1400 km transmission line from northern Manitoba, near Gillam, south to Winnipeg, including two new converter stations and ground electrodes. The transmission line will link northern hydroelectric power generation stations on the Lower Nelson River with conversion and distribution systems near Winnipeg and, in doing so, help meet growing electricity demands and improve energy security. Currently, more than 70 % of the province's electricity is transmitted via a single corridor on the Bipole I and II transmission lines. The transmission line will traverse boreal forest and caribou habitat in the north and agricultural land in the south, including several river and stream crossings along the route.

The project was subject to EA under the *Manitoba Environment Act* (Government of Manitoba 1987), with requirements set out in the project's Scoping Document (Manitoba Hydro 2010) for the assessment of the potential cumulative effects of the project in combination with other past, present, and future human actions. The CEA was conducted based on the Scoping Document and drawing also on guidance for CEA prepared under the former *Canadian Environmental Assessment Act* (e.g. Hegmann et al. 1999). The environmental impact statement concluded, as part of its CEA, that there would be no significant adverse cumulative effects caused by the project, and any residual cumulative effects following impact mitigation would be negligible (Manitoba Hydro 2011, Chap. 9).

In its final panel report on the project, the MCEC noted that CEA "should be the most important section of any environmental assessment", but went on to indicate that the MCEC has been less than satisfied with the nature of CEA conducted by proponents in Manitoba. The MCEC reported that it was "simply inconceivable—given the 50-plus-year history of Manitoba Hydro development in northern Manitoba and given that at least 35 Manitoba Hydro projects have been constructed in the north in that time—that there are few, if any, cumulative effects identified in this EIS" (MCEC 2012, p. 112).

A third-party review, commissioned by the Consumers Association of Canada (Gunn and Noble 2012), identified several deficiencies in the Bipole III CEA, including the following:

- The baseline against which cumulative effects were assessed largely ignored the cumulative effects of past hydroelectric development and other actions, and changing conditions over time.
- Few trends or condition changes were identified and analysed in the CEA, and thus there was little means to predict or model cumulative effects into the future.
- The majority of the project's potential impacts were not examined within the context of regional ecosystem health, but rather from the perspective of absorbing the project's stress.
- Much of the CEA was restricted to the proposed transmission line right-of-way and excluded other disturbances on the landscape, including the existing Bipole I and II transmission line corridors.
- The magnitude of the project's impacts was assessed against, or compared to, the effects of other actions and thus the total or cumulative effects of the project were not assessed.

Notwithstanding these deficiencies, and the noted importance of CEA, the MCEC recommended approval of the environmental licence for the Bipole III project. As a non-licensing requirement, the MCEC recommended that Manitoba Hydro, in cooperation with the Manitoba Government, conduct a regional CEA for all Manitoba Hydro projects and associated infrastructure in the Nelson River subwatershed. The transmission line is expected to be operational in 2017.

**Example 2: Spectra Energy Westcoast Connector Gas Transmission Project, British Columbia**

In 2014, Spectra Energy submitted an EA Certificate Application for the construction and operation of its Westcoast Connector Gas Transmission (WCGT) project. The WCGT project is a proposed natural gas pipeline system for the transportation of sweet gas from northeast BC to the northwest coast. The proposed pipeline route runs from the Cypress area, northwest of Fort St. John, to a proposed LNG terminal near Prince Rupert. The pipeline will be between 854 and 862 km in length, depending on the final route, and located within a 55 m right-of-way (TERA Environmental Consultants 2014). The WCGT project was subject to review under the *BC Environmental Assessment Act* (Government of BC 2002). As part of the EA Certificate Application Information Requirements, Spectra was required to undertake an assessment of the potential cumulative effects of the project for those components of the environment where potential residual effects are likely to interact with past, present, or reasonably foreseeable future projects or activities (BC EAO 2013).

The most easterly section of the pipeline is located within the Peace Region, and the traditional territory of the Blueberry River First Nation. Of concern to the Blueberry River First Nation, and to other potentially affected Aboriginal peoples whose traditional lands are traversed by the proposed pipeline, are the potential cumulative effects of the project, in combination with other industrial activities in the region, on their ability to exercise treaty rights—for example, restrictions to land and resource access, habitat fragmentation, and impacts to wildlife and fish (TERA Environmental Consultants 2014). The EA Certificate Application concludes that the project will have no significant adverse residual or cumulative environmental, economic, social, heritage, or health impacts.

The problem, however, with regard to the project's cumulative effects to traditional lands and use is that: (a) the spatial boundaries of the CEA do not capture all activities with which the project's effects may interact, cumulatively, on Blueberry River First Nation traditional lands, and (b) some of the most significant disturbances attributed to industrial activity that may interact with the proposed project affecting traditional lands and resources, namely upstream or downstream natural gas exploration and development, were not included within the scope of the assessment (TERA Environmental Consultants 2014, Table 11.19-1).

Although the project's assessment boundaries varied based on the particular affected components of concern (e.g. caribou, fish, soils), the assessment boundaries captured some, but not all, disturbances to, Blueberry River First Nation traditional lands and resources. For example, in 2011 there were 7,837 km of pipeline right-of-ways in the Beatton watershed (14,581 ha)—an area located adjacent to, and outside, the project's CEA area, but within the boundaries of traditional use of land by the Blueberry River First Nation. Lee and Hanneman (2012) report that more than half of the area (53 %) covered by oil and gas pipeline right-of-ways in the Peace Region is in the Beatton watershed—activities that may interact cumulatively with the WCGT project, but were scoped out of the assessment. Amongst the most significant industrial activities affecting BFRN traditional lands and resources are well sites—existing, decommissioned, and future development. In 2011, there were 16,211 wells in the Peace region, including 2,326 in the Upper Peace-Halfway watershed. There were an additional 8,885 wells in the Beatton watershed, with densities up to 1.5 wells per km$^2$ in some regions (Lee and Hanneman 2012)—regions outside the project's assessment area, but with the potential to act cumulatively with project impacts on Blueberry River First Nation traditional lands and resources.

The EA Certificate Application reports that there are no significant adverse cumulative impacts expected as a result of the WCGT project on habitat, species, or traditional use. However, the majority of existing industrial disturbances in the region are not considered within the spatial boundaries of the project's EA. The WCGT project is an example of a CEA with a significant geographic scope, but due to its project-based focus still fails to adequately capture cumulative environmental impacts. The project was under regulatory review at the time the chapter was written.

## Towards Better Practice

The above cases are illustrative examples of the limitations to the practice of CEA under current regulatory standards. Duinker and Greig (2006) argue that continuing the kinds and qualities of CEA currently undertaken may be doing more harm than good. Quite often, individual development projects are evaluated independently of other activities and thus deemed unlikely to cause significant adverse environmental impacts. In other cases, the magnitude of a project's impacts is erroneously measured against or compared to the effects of other projects, versus focusing on the overall impacts to those environmental components of concern. The current regulatory approach to CEA in Canada, under project-based assessment, is concerned primarily about making sure that the impacts of the project under consideration are acceptably small, or at least smaller than those of other projects, rather than assessing the total impacts of all disturbances to valued environmental components. Developers, or project proponents, are not fully to blame. Project proponents need to only do what is set out in legislation or regulation, or required of them through project-specific terms of reference.

There is a growing interest in expanding the scope of CEA from the project to the region. In the Ring of Fire, for example, a mineral resource-rich region in northern Ontario, there is pressure for the provincial government to adopt a more regional and strategic approach to addressing the cumulative effects of mineral resource development, before individual projects are approved (Chetkiewicz and Lintner 2014). Similar efforts are underway in Canada's Beaufort Sea, to address the cumulative effects of offshore energy development (Noble et al. 2013), and in the Elk Valley of BC to assess, and effectively manage, the cumulative impacts of multiple land uses, including coal-mining operations (BC MoE and BC MoFLNRO 2012). A major challenge facing such initiatives, however, is that there are few supporting legislated or regulatory provisions or mechanisms to ensure that the results of these emerging regional CEA efforts actually influence land-use and project-level decisions about resource development.

## 6.3 Synthesis

Collectively, these eight vignettes illustrated many of the challenges facing effective assessment of cumulative impacts. Vignette 1 described a situation in which multiple resource development, ranging from major dams to the ongoing activities associated with agricultural development, together, contributes to the cumulative impact profile of the watershed. This example from the Nechako Basin also highlighted the need to consider the past, present, and future of cumulative impacts particularly when put in the context of climate change. Rare are the opportunities where values are established before development and where extensive scenario planning (Chap. 3) is not only possible but expected, but the MKMA (Vignette 2) is such an area. The authors described a unique process where community planning, guided by values and supported (in principle) by legislation, defined the level and type of development that will maintain wilderness values. Although there are still many challenges to identifying and maintaining those values, the potential to minimise cumulative impacts within the MKMA contrasts starkly with the current state of the Nechako Basin and provides a point of reference for better practices in the future.

Without knowing the future, we face unintended consequences of any management action. For example, we cannot possibly manage cumulative impacts if we focus on just the decreasing value of one feature on the landscape. Instead, Vignette 3 highlighted the necessity of considering multiple, often highly interactive, values across the broader landscape. Such an approach will foster resilience in both economic and ecological systems that can benefit the human communities dependent on them, but may require a change in a business model that, for example, relies on forest diversity rather than those policies that constrain diversity.

In CEA we tend to focus on only a few species—those being either of commercial value or of high prominence. Vignette 4 provided us with an example of how an important wildlife tree is sparsely distributed in areas with relatively little resource development activity, and has no apparent utilitarian value. Whitebark pine is, however, very susceptible to the cumulative impacts of changing forest disturbance regimes (e.g. logging, mountain pine beetle, wildfire, fire suppression), invasive organisms, and climate change. Its decline has broader repercussions for subalpine landscapes and the wildlife that depend upon them. The authors challenged us to consider the unrecognised and uneconomic elements of biodiversity and ask how best to protect values that are lost through chronic, not acute, land change and the resulting cumulative impacts. Such questions, of course, can apply to a wide range of environmental values.

Landscapes include the people who live on and use those landscapes. Multiple resource development can have diverse impacts on nature-based recreation (Vignette 5), but the interaction is sometimes complex. Rather than just the simple access to nature, which is important, some development will increase recreational access, which in turn can have impacts on other resource values. Although some use the land for recreation, others derive many values and needed resources from the land—in many ways their existence. A meaningful and ultimately effective assessment framework must consider the values, knowledge, concerns, and aspirations of Aboriginal peoples (Vignette 6). Further, Aboriginal peoples are not passive recipients of the cumulative impacts of industrial development. CEA must adequately account for, and address, the broad range of cumulative effects and impacts of multiple resource development on Aboriginal peoples.

Vignette 7 highlighted the intimate relationships between people, place, identity, and health. An argument was made for a much broader scope and vision to recognise these often overlooked interconnections, and considers their implications for health and well-being in communities affected by resource development. Understanding such relationships, interconnections, and implications is made more challenging given the evolving nature of the interconnections between people, institutions, public and environmental health, and governance.

The final Vignette (8) used two examples of recent regulatory-based CEA to illustrate the constraints of the project-based model. Describing the process for approving a hydroelectric transmission line in northern Manitoba and a natural gas pipeline in northern BC, the vignette noted the limitations of project-based CEA that included ignoring past effects and impacts on the landscape, failing to project impacts into the future, and not considering potential effects and impacts from a broader perspective that was inclusive of ecosystem health. Noble argued that keeping the impacts of an individual project acceptably small may have little relation with the past, present, and future cumulative impacts for landscapes and communities.

Together these vignettes highlight the need for an appropriate scale (spatial and temporal) as a first step to revising our assessment processes. They also highlight that not only is effective planning needed, but to be successful it must represent a broad range of values that are environmental, ecological, and human centric. Allowing singular metrics such as economic worth or economic value to determine which species or which effects and impacts are given priority is to lose sight of the underlying purpose of CEA. To exclude such scope, or to limit such values, undermines and constrains the planning process, and a constrained planning process cannot effectively prepare for, or respond to, change. Collectively, the vignettes underscore the need for a new generation of integrative understanding and cumulative thinking that is informed by the past, engaged with present realities, and future oriented—an integration imperative—a vision that is developed further in Part III.

## References

Adelson, N. 2000. *Being alive well: Health and the politics of Cree wellbeing*. Toronto: University of Toronto Press.

———. 2005. The embodiment of inequity. *Canadian Journal of Public Health* 96(Suppl. no. 2): S45–S61.

Adler, R.H. 2009. Engel's biopsychosocial model is still relevant today. *Journal of Psychosomatic Research* 67: 607–611.

Alfred, T., and J. Corntassel. 2005. Being Indigenous: Resurgences against contemporary colonialism. *Government and Opposition* 40: 597–614.

AXYS Environmental Consulting. 2003. *A cumulative effects assessment and management framework (CEAMF) for Northeastern British Columbia*, vol. 1. Fort St. John: BC Oil and Gas Commission and the Muskwa-Kechika Advisory Board.

Ballard, H.L., and J.M. Belsky. 2010. Participatory action research and environmental learning: Implications for resilient forests and communities. *Environmental Education Research* 16: 611–627.

Barker, A.J. 2009. The contemporary reality of Canadian imperialism: Settler colonialism and the hybrid colonial state. *American Indian Quarterly* 33: 325–351.

Barringer, L.E., D.F. Tomback, M.B. Wunder, and S.T. McKinney. 2012. Whitebark pine stand condition, tree abundance, and cone production as predictors of visitation by Clark's nutcracker. *PLoS One* 7(5): e37663.

Barrington, D.J., S. Dobbs, and D.I. Loden. 2012. Social and environmental justice for communities of the Mekong River. *International Journal of Engineering, Social Justice, and Peace* 1: 31–49.

Barzyk, T., K. Conlon, T. Chahine, D. Hammond, V. Zartarian, and B. Schultz. 2010. Tools available to communities for conducting cumulative exposure and risk assessments. *Journal of Exposure Science and Environmental Epidemiology* 20: 371–384.

Baxter, W., W.A. Ross, and H. Spaling. 2001. Improving the practice of cumulative effects assessment in Canada. *Impact Assessment and Project Appraisal* 19: 253–262.

BC EAO (British Columbia Environmental Assessment Office). 2013. *Application Information Requirements for the proposed Northeast British Columbia to the Prince Rupert Area pipeline Project*. BC EAO: Victoria, BC.

BC Forest Practices Board. 2011. *Cumulative effects: From assessment towards management; special report.* Special Report 39. Victoria: Forest Practices Board.

BC Ministry of Forests. 1998. *The ecology of the sub-boreal spruce zone*. Victoria: BC MoF Research Branch. http://www.for.gov.bc.ca/hfd/pubs/docs/Bro/bro53.pdf. Accessed 13 Nov 2014.

BC MoE (British Columbia Ministry of Environment) and BC MoFLNRO (British Columbia Ministry of Forests Lands and Natural Resource Operations). 2012. Cumulative effects assessment framework for Natural Resource Decision Making Project Charter - Final, Version 4.0 Joint Project with Ministry of Environment and Ministry of Forests, Lands and Natural Resource Operations. http://media.wix.com/ugd/3ed831_0f72585af7574726a5478ba3bda2c65b.pdf. Accessed 20 Jan 2015.

BC MoFLNRO (British Columbia Ministry of Forests Lands and Natural Resource Operations. 2012. *Mid-term timber supply: Lakes Timber Supply Area*. Victoria: BC Government. http://www.for.gov.bc.ca/hfp/mountain_pine_beetle/mid-term-timber-supply-project/Lakes%20TSA.pdf. Accessed 14 Nov 2014.

BC Oil and Gas Commission. 2014. Are the visual quality objectives, as defined under FRPA, applicable to OGAA applications? https://www.bcogc.ca/are-visual-quality-objectives-defined-under-frpa-applicable-ogaa-applications. Accessed 13 Nov 2014.

BC Wilderness Tourism Association. 2013. Value of wilderness tourism. http://www.wilderness-tourism.bc.ca/value.html. Accessed 13 Nov 2014.

Benke, A.C., and C.E. Cushing (eds.). 2005. *Rivers of North America*. New York: Elsevier Academic Press.

Benusic, M. 2013. Fracking in BC: A public health concern. *British Columbia Medical Journal* 55: 238–239.

Berry, H.L., J.R. Butler, C.P. Burgess, U.G. King, K. Tsey, Y.L. Cadet-James, C.W. Rigby, and B. Raphael. 2010. Mind, body, spirit: Co-benefits for mental health from climate change adaptation and caring for country in remote Aboriginal Australian communities. *New South Wales Public Health Bulletin* 21: 139–145.

Best, A., and B. Holmes. 2010. Systems thinking, knowledge and action: Towards better models and methods. *Evidence and Policy: A Journal of Research, Debate and Practice* 6: 145–159.

Booth, A.L., and N.W. Skelton. 2011. "You spoil everything!" Indigenous peoples and the consequences of industrial development in British Columbia. *Environment, Development and Sustainability* 13: 685–702.

Booth-Sweeny, L. 2009. *Living stories about living systems: Connected wisdom*. Schlumberger Excellence in Educational Development (SEED). http://www.planetseed.com/home.

Boyle, M., and H. Dowlatabadi. 2011. Anticipatory adaptation in marginalized communities within developed countries. *Advances in Global Change Research* 42: 461–473.

Brown, P., S. Zavestoski, S. McCormick, B. Mayer, R. Morello-Frosch, and R. Gasior Altman. 2004. Embodied health movements: New approaches to social movements in health. *Sociology of Health and Illness* 26: 50–80.

Brown, V.A., J.A. Harris, and J.Y. Russell. 2010. *Tackling wicked problems through the transdisciplinary imagination*. London: Earthscan.

Budruk, M., and S.A. Wilhelm Stanis. 2013. Place attachment and recreation experience preference: A further exploration of the relationship. *Journal of Outdoor Recreation and Tourism* 1–2: 51–61.

Bunch, M.J., K.E. Morrison, M. Parkes, and H.D. Venema. 2011. Promoting health and wellbeing by managing for social–ecological resilience: The potential of integrating ecohealth and water resources management approaches. *Ecology and Society* 16(1): 6.

Butler, R.W. 1990. Alternative tourism: Pious hope or Trojan Horse? *Journal of Travel Research* 28(3): 40–45.

Canada Parks Council. 2011. Benefits of Parks and Protected Areas. http://www.parks-parcs.ca/english/cpc/benefits.php. Accessed 19 Dec 2013.

Chandler, M.J., and C.E. Lalonde. 2004. Transferring whose knowledge? Exchanging whose best practices? On knowing about Indigenous knowledge and Aboriginal suicide. In *Aboriginal policy research volume II: Setting the agenda for change*, ed. J.P. White, P. Maxim, and D. Beavon, 111–123. Toronto: Thompson Educational Press.

Cheng, A.S., L.E. Kruger, and S.E. Daniels. 2003. "Place" as an integrating concept in natural resource politics: Propositions for a social science research agenda. *Society and Natural Resources* 16: 87–104.

Chetkiewicz, C., and A.M. Lintner. 2014. *Getting it right in Ontario's north: The need for regional strategic environmental assessment in the Ring of Fire [Wawangajing]*. Toronto: Wildlife Conservation Society and Ecojustice.

Dubé, M., and J. Wilson. 2013. Accumulated state assessment of the Peace-Athabasca-Slave River system. *Integrated Environmental Assessment and Management* 9: 405–425.

Duinker, P.N., and L.A. Greig. 2006. The impotence of cumulative effects assessment in Canada: Ailments and ideas for redeployment. *Environmental Management* 37: 153–161.

FBC (Fraser Basin Council). 2012. *Identifying health concerns relating to oil and gas development in Northeastern BC: Human Health Risk Assessment—Phase 1 Report*. Victoria: BC Ministry of Health. http://www.health.gov.bc.ca/protect/oil-gas-assessment.html. Accessed 9 Nov 2014.

First Nations Environmental Health Innovation Network. 2008. Human environmental health impact assessment: A framework for Indigenous communities. http://www.fnehin.ca/uploads/docs/Project_Overview.pdf. Accessed 13 Nov 2014.

Fisher, P. 2008. Wellbeing and empowerment: The importance of recognition. *Sociology of Health and Illness* 30: 583–598.

Foucault, M. 2003. *The birth of the clinic*. London: Routledge.
Gartner, W.C., and D.W. Lime. 2000. The big picture: A synopsis of contributions. In *Trends in outdoor recreation, leisure, and tourism*, ed. W.C. Gartner and W.C. Lime, 1–14. Cambridge, MA: CAB International.
Gies, E. 2006. *The health benefits of parks: How parks help keep Americans and their communities fit and healthy*. San Francisco: The Trust for Public Land.
Government of BC. 1998. *Muskwa-Kechika Management Area Act, 1998*. Bill 37. http://leg.bc.ca/36th3rd/3rd_read/gov37-3.htm. Accessed 11 Jan 2015.
———. 2002. *Environmental Assessment Act, 2002* (S.B.C. 2002, C. 43). http://www.bclaws.ca/civix/document/id/complete/statreg/02043._01 Accessed 10 Dec 2014.
———. 2011a. *The social and economic impacts of BC recreation sites and trails*. Victoria: Ministry of Forests, Lands and Natural Resource Operations.
Government of Canada. 2002. *Species at Risk Act, 2002* (S.C. 2002, C. 29). http://laws-lois.justice.gc.ca/eng/acts/S-15.3/. Accessed 29 Mar 2015.
———. 2012. *Canadian Environmental Assessment Act, 2012* (S.C. 2012, C. 19, S. 52). http://laws-lois.justice.gc.ca/eng/acts/c-15.21/page-1.html. Accessed 12 Dec 2014.
Government of Manitoba. 1987. *Manitoba Environmental Act, 1987* (C.C.S.M. 1987, C. E125). https://web2.gov.mb.ca/laws/statutes/ccsm/e125e.php. Accessed 6 Apr 2015.
Gracey, M., and M. King. 2009. Indigenous health part 1: Determinants and disease patterns. *The Lancet* 374: 65–75.
Greenwood, M., and J. Place. 2009. Executive summary: The health of First Nations, Inuit and Metis Children in Canada. In *Aboriginal chlidren's health: Leaving no child behind. Canadian Supplement to The State of the World's Children*, ed. NCCAH, 1–10. Toronto: UNICEF Canada.
Gunn, J., and B.F. Noble. 2012. *Critical review of the cumulative effects assessment undertaken by Manitoba Hydro for the Bipole III project*. Winnipeg: Prepared for the Public Interest Law Centre.
Gustine, D.D., K.L. Parker, R.J. Lay, M.P. Gillingham, and D.C. Heard. 2006. Calf survival of woodland caribou in a multi-predator ecosystem. *Wildlife Monographs* 165: 1–32.
Hall, R. 2012. Diamond mining in Canada's Northwest Territories: A colonial continuity. *Antipode* 45: 376–393.
Halpenny, E.A. 2010. Pro-environmental behaviours and park visitors: The effect of place attachment. *Journal of Environmental Psychology* 30: 409–421.
Hammitt, W.E., E.A. Backlund, and R.D. Bixler. 2006. Place bonding for recreation places: Conceptual and empirical development. *Leisure Studies* 25: 17–41.
Harris, C. 2004. How did colonialism dispossess? Comments from an edge of empire. *Annals of the Association of American Geographers* 94: 165–182.
Hartman, G.F. 1996. Impacts of growth in resource use and human population on the Nechako River: A major tributary of the Fraser River, British Columbia, Canada. *Geojournal* 40: 147–164.
Hegmann, G., C. Cocklin, R. Creasey, S. Dupuis, A. Kennedy, L. Kingsley, W. Ross, H. Spaling, and D. Stalker. 1999. *Cumulative effects assessment practitioners guide*. Hull: AXYS Environmental Consulting, and the CEA Working Group for the Canadian Environmental Assessment Agency.
Heinemeyer, K., R. Tingey, K. Ciruna, T. Lind, J. Pollock, B. Butterfield, J. Griggs, P. Iachetti, C. Bode, T. Olenicki, E. Parkinson, C. Rumsey, and D. Sizemore. 2004a. *Conservation area design for the Muskwa-Kechika Management Area. Volume 1: Final report*. Prepared for the British Columbia Ministry of Sustainable Resource Management.
———. 2004b. *Conservation area design for the Muskwa-Kechika Management Area. Volume 2: Appendices*. Prepared for the British Columbia Ministry of Sustainable Resource Management.
Hoekstra, G. 2014. Mount Polley mine tailings spill nearly 70 per cent bigger than first estimated. *Vancouver Sun*. September 3. http://www.vancouversun.com/Mount+Polley+mine+tailings+spill+nearly+cent+bigger+than+first+estimated/10172302/story.html . Accessed 14 Nov 2014.
Hunt, L.M., R.H. Lemelin, and K.C. Saunders. 2009. Managing forest road access on public lands: A conceptual model of conflict. *Society and Natural Resources* 22: 128–142.

IHRC (International Human Rights Clinic). 2010. *Bearing the burden: The effects of mining on First Nations in British Columbia*. Cambridge: Harvard Law School. http://www.ceaa-acee.gc.ca/050/documents/p63928/92021E.pdf. Accessed 14 Nov 2014.

IPCC (Intergovernmental Panel on Climate Change). 2013. *Climate change 2013: The physical science basis*. Cambridge: Cambridge University Press.

IPIECA (International Petroleum Industry Environmental Conservation Association). 2005. A guide to health impacts assessment in the oil and gas industry. http://www.ipieca.org/publication/health-impact-assessments. Accessed 14 Nov 2014.

Keskinen, M., M. Kummu, M. Käkönen, and O. Varis. 2012. Mekong at the crossroads: Next steps for impact assessment of large dams. *AMBIO: A Journal of the Human Environment* 41: 319–324.

King, M., A. Smith, and M. Gracey. 2009. Indigenous health part 2: The underlying causes of the health gap. *The Lancet* 374: 76–85.

Kingsley, J., M. Townsend, and C. Henderson-Wilson. 2013. Exploring Aboriginal people's connection to country to strengthen human-nature theoretical perspectives. In *Ecological health: Society, ecology and health, Volume 15*, ed. M.K. Gislason, 936–944. Bingley: Emerald Group Publishing.

Krieger, N. 2012. Methods for the scientific study of discrimination and health: An ecosocial approach. *American Journal of Public Health* 102: 936–944.

Kruger, L.E. 2006. Recreation as a path for place making and community building. *Leisure* 30: 383–392.

Kuo, F.E.M. 2009. *Parks and other green environments: essential components of a healthy human habitat*. Ashburn: National Recreation and Park Association.

Kurz, W.A., C.C. Dymond, G. Stinson, G.J. Rampley, E.T. Neilson, A.L. Carroll, T. Ebata, and L. Safranyik. 2008. Mountain pine beetle and forest carbon feedback to climate change. *Nature* 452: 987–990.

LaDuke, W. 1999. *All our relations: Native struggle for lands and life*. London: Zed Books.

Lee, K. 2010. How do we move forward on the social determinants of health: The global governance challenges. *Critical Public Health* 20: 5–14.

Lee, P., and M. Hanneman. 2012. *Atlas of land cover, industrial land uses and industrial-caused land change in the peace region of British Columbia*. Edmonton: Global Forest Watch Canada.

Leung, Y.-F., J.L. Marion, and T.A. Farrell. 2008. Recreation ecology in sustainble tourism and ecotourism: A strengthening role. In *Tourism, recreation, and sustainability: Linking culture and the environment*, ed. S.F. McCool and R.N. Moisey, 19–36. New York: CABI Publishing.

Mah, S., and K. Astridge. 2014. *Landscape-level ecological tree species benchmarks pilot project: First approximation benchmarks in five British Columbia Timber Supply Areas*. Victoria: B.C. Ministry of Forests, Lands and Natural Resource Operations.

Manion, P.D. 1981. *Tree disease concepts*. Englewood Cliffs: Prentice-Hall.

Manitoba Hydro. 2010. *The Bipole III transmission project environmental assessment scoping document*. Winnipeg: Manitoba Hydro.

———. 2011. *Bipole III transmission project: A major reliability initiative*. Environmental Impact Statement, pursuant to The Manitoba Environment Act. Winnipeg: Manitoba Hydro.

MCEC (Manitoba Clean Environment Commission). 2012. *Bipole III transmission project: Report on public hearing*. Winnipeg: Manitoba Clean Environment Commission.

McLane, S.C., and S.N. Aitken. 2012. Whitebark pine (*Pinus albicaulis*) assisted migration potential: Testing establishment north of the species range. *Ecological Applications* 22: 142–153.

Meyers Norris Penny LLP. 2011. *The social and economic impacts of BC recreation sites and trails*. Prepared for the Ministry of Forests, Lands and Natural Resource Operations.

Milakovic, B., and K.L. Parker. 2011. Using stable isotopes to define diets of wolves in northern British Columbia, Canada. *Journal of Mammalogy* 92: 295–304.

Millennium Ecosystem Assessment. 2005. *Ecosystems and human wellbeing*, vol. 5. Geneva: World Health Organization. http://www.who.int/entity/globalchange/ecosystems/ecosys.pdf. Accessed 14 Nov 2014.

Mueller-Dombois, D. 1988. Towards a unifying theory for stand-level dieback. *Geojournal* 17: 249–251.
Nielsen, J., B.F. Noble, and M. Hill. 2012. Wetland assessment and impact mitigation decision support framework for linear development projects: The Louis Riel Trail, Highway 11 North project, Saskatchewan, Canada. *The Canadian Geographer* 56: 117–139.
Noble, B.F., S. Ketilson, A. Aitken, and G. Poelzer. 2013. Strategic environmental assessment opportunities and risks for Arctic offshore energy planning and development. *Marine Policy* 39: 296–302.
NRCAN (Natural Resources Canada). 2013. Responsible resource development and related legislative, regulatory and policy improvements to modernize the regulatory system for project reviews. http://www.nrcan.gc.ca/environmental-assessment/149. Accessed 14 Nov 2014.
O'Rourke, D., and S. Connolly. 2003. Just oil? The distribution of environmental and social impacts of oil production and consumption. *Annual Review of Environmental Resources* 28: 587–617.
OCMOH (Office of the Chief Medical Health Officer). 2012. *Chief Medical Officer of Health's Recommendations Concerning Shale Gas Development in New Brunswick*. Fredericton: OCMOH. http://sustainabilityresearch.wp.rpi.edu/files/2013/04/NewBrunswickFrackingReportSept2012.pdf. Accessed 14 Nov 2014.
Oh, C.-O., S.O. Lyu, and W.E. Hammitt. 2012. Predictive linkages between recreation specialization and place attachment. *Journal of Leisure Research* 44: 70–87.
Parkes, M.W. 2011. *Ecohealth and watersheds in northern BC: Improving social and environmental determinants of health through integrated health governance; executive summary*. Prince George: University of Northern British Columbia. http://www.unbc.ca/sites/default/files/sections/parkes/2011-061-3executivesummary.pdf. Accessed 18 Jan 2014.
Parkes, M.W., K.E. Morrison, M.J. Bunch, L.K. Hallstrom, R.C. Neudoerffer, H.D. Venema, and D. Waltner-Toews. 2010. Towards integrated governance for water, health and social–ecological systems: The watershed governance prism. *Global Environmental Change* 20: 693–704.
Parkes, M.W., S. De Leeuw, and M.A. Greenwood. 2011. Warming up to the embodied context of First Nations health: A critical intervention into and analysis of health and climate change research. *International Public Health Journal* 2: 477–485.
Parkins, J.R. 2011. Deliberative democracy, institution building, and the pragmatics of cumulative effects assessment. *Ecology and Society* 16(3): 20.
Parlee, B., J. O'Neil, and Lutsel K'e Dene First Nation. 2007. The Dene way of life: Perspectives on health from Canada's north. *Journal of Canadian Studies* 41: 112–133.
Peterson, G.D., T.D. Beard Jr., B.E. Beisner, E.M. Bennett, S.R. Carpenter, G.S. Cumming, and T.D. Havlicek. 2003. Assessing future ecosystem services: A case study of the Northern Highlands Lake District, Wisconsin. *Conservation Ecology* 7(3): 1.
Picketts, I.M., J. Curry, S.J. Déry, and S.J. Cohen. 2013. Learning with practitioners: Climate change adaptation priorities in a Canadian community. *Climatic Change* 118: 321–337.
Place, J., and N. Hanlon. 2011. Kill the lake? Kill the proposal: Accommodating First Nations' environmental values as a first step on the road to wellness. *GeoJournal* 76: 163–175.
Poland, B. 2010. The transition handbook: From oil dependency to local resilience, by Rob Hopkins. *Critical Public Health* 20: 385–387.
Reading, C.L., and F. Wien. 2009. *Health inequalities and the social determinants of Aboriginal Peoples' health*. Prince George: National Collaborating Centre for Aboriginal Health.
Richmond, C. 2015. The relatedness of people, land and health: Stories from Anishinabe Elders. In *Determinants of Indigenous peoples' health in Canada: Beyond the social*, ed. M. Greenwood, S. de Leeuw, N. Lindsay, and C. Reading. Toronto: Canadian Scholars Press.
Richmond, C., S.J. Elliott, R. Matthews, and B. Elliott. 2005. The political ecology of health: Perceptions of environment, economy, health and wellbeing among 'Namgis First Nation'. *Health and Place* 11: 349–365.
Richmond, C.A., and N.A. Ross. 2009. The determinants of First Nation and Inuit health: A critical population health approach. *Health and Place* 15: 403–411.

Rodenhuis, D.R., K.E. Bennett, A.T. Werner, D. Bronaugh, and T.Q. Murdock. 2009. *Hydro-climatology and future climate impacts in British Columbia*. Victoria: Pacific Climate Impacts Consortium.

Salmo Consulting, and Diversified Environmental Services. 2003. CEAMF study: *Volume 2; Cumulative indicators, thresholds, and case studies*. Calgary: Prepared for the BC Oil and Gas Commission, and the Muskwa-Kechika Advisory Board.

Sanderson, D., I.M. Picketts, S.J. Déry, B. Fell, S. Baker, E. Lee-Johnson, and M. Auger. 2015. Climate change and water at Stellat'en First Nation, British Columbia, Canada: Insights from western science and traditional knowledge. *The Canadian Geographer* 59: 136–150.

Sauve, M. 2007. Canadian dispatches from medical fronts: Fort McMurray. *Canadian Medical Association Journal* 117: 26.

Schmidt, C. 2011. Blind rush? Shale gas boom proceeds amid human health questions. *Environmental Health Perspectives* 119: A349–A353.

Seitz, N., C. Westbrook, and B. Noble. 2011. Bringing science into river systems cumulative effects assessment practice. *Environmental Impact Assessment Review* 31: 180–186.

Sexton, K. 2012. Cumulative risk assessment: An overview of methodological approaches for evaluating combined health effects from exposure to multiple environmental stressors. *International Journal of Environmental Research and Public Health* 9: 370–390.

Sheppard, S.R.J., A. Shaw, D. Flanders, S. Burch, A. Wiek, J. Carmichael, J. Robinson, and S. Cohen. 2011. Future visioning of local climate change: A framework for community engagement and planning with scenarios and visualisation. *Futures* 43: 400–412.

Shrestha, R.R., M.A. Schnorbus, A.T. Werner, and A.J. Berland. 2012. Modelling spatial and temporal variability of hydrologic impacts of climate change in the Fraser River basin, British Columbia, Canada. *Hydrological Processes* 26: 1840–1860.

Shultis, J., and R. Rutledge. 2003. The Muskwa-Kechika Management Area: A model for sustainable development of wilderness. *International Journal of Wilderness* 9: 12–17.

Stephen, C. 2013. Towards a new definition of animal health: Lessons learned from the Cohen Commission and the SPS agreement. *Optimum Online: The Journal of Public Sector Management* 43: 1–5.

Stephens, T. 1988. Physical activity and mental health in the United States and Canada: Evidence from four population surveys. *Preventive Medicine* 17: 35–47.

Struzik, E. 2013. *Underground intelligence: The need to map, monitor, and manage Canada's groundwater resources in an era of drought and climate change*. Toronto: University of Toronto Munk School of Global Affairs Program on Water Issues.

Taylor, S., and A. Carroll. 2003. Disturbance, forest age, and mountain pine beetle outbreak dynamics in BC: A historical perspective. In *Mountain Pine Beetle Symposium: Challenges and solutions*, ed. T. Shore, J. Brooks, and J. Stone, 41–51. Victoria: Pacific Forestry Centre Information report No. 399.

TERA Environmental Consultants. 2014. *Environmental Assessment Certificate Application for the Westcoast Connector Gas Transmission Project. Prepared for review under the British Columbia Environmental Assessment Act*. Vancouver: Prepared for Westcoast Connector gas Transmission Ltd.

Tomback, D., S.F. Arno, and R.E. Keane (eds.). 2001. *Whitebark Pine Communities: Ecology and restoration*. Washington: Island Press.

Tuck, E., and K.W. Yang. 2012. Decolonization is not a metaphor. *Decolonization: Indigeneity, Education and Society* 1(1): 1–45.

Waggoner, P.E. 1962. Weather, space, time, and chance of infection. *Phytopathology* 52: 1100–1108.

Warburton, D.E.R., C.W. Nicol, and S.S.D. Bredin. 2006. Health benefits of physical activity: The evidence. *Canadian Medical Association Journal* 174: 801–809.

West, P., J. Igoe, and D. Brockington. 2006. Parks and peoples: The social impact of protected areas. *Annual Review of Anthropology* 35: 251–277.

WHO (World Health Organization). 1986. *Ottawa charter for health promotion*. In: Milestones in health promotion. Statements from Global Conferences. WHO/NMH/CHP/09.01 Geneva:

World Health Organisation. http://www.who.int/healthpromotion/conferences/previous/ottawa/en/. Accessed 9 Nov 2014.
Williamson, T.B., T.B. Price, J.L. Beverley, P.M. Bothwell, B. Frenkel, J. Park, and M.N. Patriquin. 2008. *Assessing potential biophysical and socioeconomic impacts of climate change on forest-based communities: A methodological case study*. Edmonton: Natural Resources Canada. http://publications.gc.ca/collections/collection_2009/nrcan/Fo133-1-415E.pdf. Accessed 14 Nov 2014.
Wilson, K. 2003. Therapeutic landscapes and First Nations peoples: An exploration of culture, health and place. *Health and Place* 9: 83–93.
Windsor, J.E., and J.A. McVey. 2006. Annihilation of both place and sense of place: The experience of the Cheslatta T'En Canadian First Nation within the context of large-scale environmental projects. *Geographical Journal* 171: 146–166.
Woods, A., and K.D. Coates. 2013. Are biotic disturbance agents challenging basic tenets of growth and yield and sustainable forest management? *Forestry* 86: 543–554.
Woods, A., K.D. Coates, and A. Hamann. 2005. Is an unprecedented dothistroma needle blight epidemic related to climate change? *BioScience* 55: 761–769.
Yellowstone to Yukon Conservation Initiative. 2012. *Muskwa-Kechika Management Area Biodiversity Conservation and Climate Change Assessment, Summary Report*. Canmore: Prepared for the Muskwa-Kechika Advisory Board.
Yuct Ne Senxiymetkwe Camp's Facebook page. October 26 2014. https://www.facebook.com/yuctnesenxiymetkwecamp/timeline. Accessed 1 Nov 2014.

# Part IV
# Synthesis

# Chapter 7
# An Imperative for Change: Towards an Integrative Understanding

**Margot W. Parkes, Chris J. Johnson, Greg R. Halseth, and Michael P. Gillingham**

## 7.1 Introduction

Concern about cumulative effects and impacts is not new. In keeping with the key terms introduced in Box 1.1 (in Chap. 1), preceding chapters have described the growing need to recognise that the *effects* (changes) caused by natural resource development can have a range of consequences, resulting in both anticipated and unanticipated *impacts* on the environment, communities, and health. Efforts to address such impacts have developed and evolved over the last 40 years. As detailed in Chaps. 2–5, much of the initial effort was focused on EIA (Noble 2010), which was followed by increased attention to SIA (Yukon Government 2002; Vanclay and Esteves 2011), and most recently HIA (Wernham 2011; Kinnear et al. 2013). Considerable public policy and scientific research have been developed within and around these assessment frameworks to aid in decision-making and mitigation of project impacts. Concurrent with the evolution of processes to evaluate individual projects was the formal recognition and development of methods and processes to assess and manage cumulative impacts (Hegmann et al. 1999; Kenna 2011). Although cumulative effects have been a consideration in policy and regulatory processes since the 1970s and 1980s in both the USA and Canada, there is very little formal recognition that a comprehensive understanding of cumulative impacts will demand attention to the environmental, social, and health impact dynamics outlined in the previous chapters.

Despite progress in the understanding, research, and measurement of cumulative impacts, there have also been increasing numbers of *cracks* apparent in both processes and outcomes. First, as detailed earlier in this book, these include concerns about the limited spatial and temporal scale of assessment. A second concern is the

M.W. Parkes (✉) • C.J. Johnson • G.R. Halseth • M.P. Gillingham
The University of Northern British Columbia, Prince George, BC, Canada V2N 4Z9
e-mail: Margot.Parkes@unbc.ca; Chris.Johnson@unbc.ca; Greg.Halseth@unbc.ca; Michael.Gillingham@unbc.ca


M.P. Gillingham et al. (eds.), *The Integration Imperative*,
DOI 10.1007/978-3-319-22123-6_7

inadequacies that arise when environmental, social, and health impact assessment approaches are translated into regulatory or legislative frameworks that become highly prescribed, cumbersome, and increasingly costly and legalised. Experience shows some processes becoming year-long hearings focused on rotating suites of lawyers calling various expert witnesses. What is lost in such an approach is an effort to build true understanding and share information towards better decision-making. A third crack in our current approach to understanding the cumulative impacts of natural resource development has to do with the increasing focus only on large mega-projects to the exclusion of the many small projects and activities that occur much more routinely, as exemplified in earlier chapters. Recent changes in EA legislation in Canada suggest that governments are less interested in regulating resource development projects with a limited spatial or temporal scope and magnitude that are assumed to have no significant adverse environmental impacts. Thus, there are now fewer tools and less attention focused on addressing cumulative impacts that result from the many small changes that such development projects create in human and ecological communities.

Despite a considerable period of experimentation and evolution in process, our current approaches to impact assessment (spanning environmental, social, and health dynamics) are all found to be wanting, especially in relation to cumulative concerns. It is clear that progress towards comprehensive understanding will not arise from fixing and making minor amendments to existing mechanisms. Rather, moving forward with a better understanding of cumulative environmental, community, and health impacts of natural resource development will demand innovation and integration that surpasses what is possible within existing project-specific approaches.

The first hurdle we must overcome in reorganising our understanding and approach to cumulative impacts is to change the mindset of people who are *stuck* in the current regulatory and approval processes. We remind people that the current approaches did arise from somewhere, and in response to something. Although current approaches for the assessment and management of cumulative impacts are now outdated and ineffective, past experience tells us that dramatic and fundamental change in our approach is possible.

In the following sections we place our contemporary challenges in the context of other notable and historic points of change in how impact assessment is framed. We focus particularly on one example: the Mackenzie Valley Pipeline Inquiry headed by Justice Thomas Berger in the 1970s (Berger 1977a, b), which proved to be a defining moment in the development and evolution of EA processes in Canada and also had far-reaching international implications. Although these points of reference occurred prior to the contemporary EA legislative context, our argument is that these precedents provide important reminders of the potential—and imperative—for change when existing processes are inadequate. Just as the challenges posed by the Berger Inquiry resulted in a fundamentally new approach to environmental and social impact assessment, so too do we need now to respond to the challenges of better understanding and dealing with cumulative impacts from natural resource development projects.

To set the stage for responding to the challenge of building a better way to understand impacts, this chapter is divided into six sections, each examining different ways in which the complex historical and contemporary context of impact assessment can be seen as part of larger social–ecological systems with an inherent capacity—and need—for adaptation and change (Berkes et al. 2003). Following this introduction, the second section provides historical background to the development of our current environmental impact, social impact, and health impact assessment processes. We begin with a look at the landmark Northern Frontier-Northern Homeland reports from the Berger Inquiry as a precedent for the possibility of undertaking profound shifts in public and natural resource policy. The third section explores the pressures that have been transforming our society more generally in the past 40 years since the Berger Inquiry report. These pressures provide a backdrop for the significant challenges facing current social, health, and environmental impact assessment processes that limit our ability to effectively understand more fully the cumulative impacts of natural resource development. This broader social, political, and economic context is essential for understanding the characteristics of our current systems, including the potential and imperatives for change. The fourth section builds on this historical context to revisit the evolution of EA, including how these approaches have been codified by legislation. We explore the ways by which such frameworks have become increasingly mechanistic and quasi-judicial limiting their capacity to respond to the full range of cumulative impacts. The fifth section builds from the limitations of the current project-specific regulatory system to revisit the imperative for change created by the complex integrative, regional, and cumulative dynamics of natural resource development across social, health, and environmental systems.

Expanding on themes introduced in Chap. 1, we argue that cumulative impacts are wicked problems (Brown et al. 2010) that will not be addressed by tinkering with our current system. Rather—informed by precedents that dramatic changes in approach are possible—we are reminded of the need for a revolutionary progression that is guided by lessons from the past, but is premised on bold visions for the future and new ways of understanding our interactions and dependence on the natural world.

## 7.2 Precedents for Change: Current Challenges in Historical Context

As reviewed in Chap. 2, the formal acknowledgement of cumulative impacts began more than 30 years ago (Peterson et al. 1987). In the USA, CEA was always intended to be one of the central aspects of the *National Environmental Policy Act* (United States Federal Government 1969), but regulations specifically emphasizing cumulative effects were not written into that Act until 1978 (Schultz 2012). In Canada, cumulative effects entered the policy discussion in the mid-1980s, but did not become legislated until 1995, when the *Canadian Environmental Assessment Act* (Government of Canada 1992) required consideration of cumulative effects within all EAs (see Duinker et al. 2012; Chap. 2).

When looking for precedents of change in our approach to effects and impact assessment, the processes influencing the inclusion of social impacts into EAs are instructive. These impacts were first formalised in the US *National Environmental Policy Act* in 1969 (United States Federal Government 1969) and gained profile in Canada in 1973 in the context of changes to Inuit culture related to the Alaska pipeline (Burdge and Vanclay 1995). Soon thereafter, the Berger Inquiry created a dramatic shift in the culture of impact assessment in Canada—representing the first case in which social impacts were given consideration and small resource-dependent communities were offered a voice in project decision-making (Berger 1977; Burdge and Vanclay 1995). This game-changing precedent not only reminds us that change can happen, but also reaffirms the ongoing potential for necessary change within current systems, despite their constraints.

The Mackenzie Valley Pipeline Inquiry, also known as the Berger Inquiry after Justice Thomas Berger, was established by Canadian Parliament in March 1974 and was tasked with providing a review of the environmental, social, and economic impacts of a proposed natural gas pipeline. Of particular note was the ways in which the Inquiry (Berger 1977a, b) increased the level of inclusion of all interests while at the same time relating the economics of a massive industrial resource project to a more local scale. The report considered the views of local people as being of central importance to understanding impacts:

> I discovered that people in the North have strong feelings about the pipeline and large-scale frontier development. I listened to a brief by northern businessmen in Yellowknife who favour a pipeline through the North. Later, in a native village far away, I heard virtually the whole community express vehement opposition to such a pipeline. Both were talking about the same pipeline; both were talking about the same region—but for one group it is a frontier, for the other a homeland. (Berger 1977, p. vii)

These insights were evident in the title of the report *Northern Frontier, Northern Homeland,* which recognised that the landscapes affected by the pipeline were the social and cultural home of large numbers of people. The relative success of Aboriginal peoples in asserting their visions of northern development facilitated by the Berger Inquiry also led to a major change in relations between Aboriginal peoples and non-Aboriginal peoples around resource development in Canada (see also Sect. 1.7 for context and terminology). Sabin (1995) described how the Mackenzie Valley Pipeline inquiry sought the engagement of diverse viewpoints especially through the process of community hearings in the north. The Berger Inquiry, and its outcomes, changed the form of government-public dialogue and resulting policy and legislation that guided natural resource decision-making.

Consistent with the thesis and ultimate conclusions of this book (Chap. 8), Berger took an unexpected and unprecedented position in the Mackenzie Valley Pipeline inquiry. The innovations of the Berger Inquiry can be seen to herald a major change in the paradigm of how individual projects were assessed. It set the model for future inquires. For example, in BC the Review Panel appointed by the BC Utilities Commission in the early 1990s to conduct hearings on the Kemano Completion Project proposed by Alcan followed very closely Berger's framework of holding formal hearings on technical and legal matters as well as going out to affected communities to hear from all stakeholders on the meanings and implications of the proposal on their lives and livelihoods (BC Utilities Commission 1994). In addition to offering a precedent for future change, other key elements of the Berger example relevant to our current challenges were his orientation to regional insights and recognition of the value of articulating a regional vision of sustainability to guide decision-making. Following the Berger Inquiry, there was no going back to the previous model, and as such this event represented a turning point in how impacts were understood, whose voices were heard, how we should balance different interests and imperatives, how we should prepare to mitigate those elements that can be mitigated and which dynamics were considered relevant.

We propose that it is possible to see the current challenges with cumulative impact assessment as warranting a turning point—and paradigm shift—that share some of the characteristics of change arising in the aftermath of the Berger Inquiry. Prior to the Berger Inquiry, environmental impacts were viewed as a technical problem to be solved by project proponents; this approach proved inadequate to the deeply contextual and contested nature of the issues associated with the Mackenzie pipeline. After the Berger Inquiry, the value of viewing impact assessments in context was recognised, and previously unacknowledged considerations were taken into account. Although not necessarily a shift towards cumulative impacts, the Berger Inquiry is especially instructive as an example of leadership that identified new and innovative paths forward when existing processes were found wanting.

## 7.3 Related Societal Shifts: Further Precedents and Prompts for Change

We argue that understanding the current context and calls for change in relation to cumulative environmental, community, and health impacts from natural resource development demands an awareness of precedents and an ability to understand future change within a wider social and historical context. In particular the evolution, and limitations, of current processes and legislation needs to be seen within the wider context of social, economic, and political restructuring. This restructuring, accelerated after the global economic recession of the early 1980s, has changed the way we approach and execute dialogue and debate about cumulative impacts. In this section, we explore this background process of transformation through restructuring. The topics include the transition from a Keynesian political economy framework to a neoliberal political economy framework, and the implications of this transition for government, industry, communities, and regions.

At its most basic, the 1980s marked a pivot point in the transition from a Keynesian to a neoliberal political economy framework. As a response to the excesses of the free market, and the catastrophe of the Great Depression, a Keynesian framework advocated a strong role for government and public policy in supporting and regulating economic activity (Tonts and Jones 1997; Scarpa 2013; Sullivan et al. 2014). Such a public policy approach focused on "continual state intervention to manage the contradictions of capitalism to the benefit of the nation and its least well-off citizens" (Sheppard 2009, p. 548). Regulation and oversight were designed so that the processes of supply and demand in setting market valuations could operate, but could not become too skewed or unbalanced. Similarly, the state would intervene monetarily to support the economy when needed to limit the harmful negative tendencies that acted out during recessions and depressions. In addition, public policy intervention in social programs assisted capital by removing some of the costs associated with the delivery of social welfare services such as education and health care that were so vital to industrial growth.

The waning of the Keynesian framework began during the economic boom years of the 1970s (Harvey 2005). Capital began to argue for greater freedom to pursue opportunities for profit. With support from some schools of economic study, and under conservative leaders such as Margaret Thatcher in Britain and Ronald Reagan in the USA, public policy began a shift towards a more neoliberal political economy framework. In simple terms, a neoliberal framework assumes that only an unfettered market can correctly value goods and services (Peck and Tickell 2002; Tonts and Haslam-McKenzie 2005; Markey et al. 2008; Shortall and Warner 2010). To start, the transition towards neoliberalism entails deregulation and public policy pullback from economic decision-making. This is accompanied by a reduction of the tax burden on capital, either by directly reducing tax and royalty regimes or by reducing the overall rate of growth of government spending. When coupled with debt and deficit reduction, this results in a steady reduction in the capacity of government to monitor or plan for economic change. In the context of natural resource management, both at strategic and operational scales, self-regulation by industry of its own (generally confined to environmental) impacts becomes more common. Neoliberal processes reduce government intervention through deregulation and rely on market benefits to influence corporate environmental and social responsibility (Higgins and Lockie 2002; Castree 2008). Certification of forest products and practices is an example of how neoliberal policy and the markets might increase or ensure sustainable natural resource practices. There is still debate, however, about the efficacy of such processes, especially where sustainability is not valued by the market, certification standards lack rigour and inclusivity, and the public has a low level of trust in industries that have the power to manipulate and significantly degrade ecosystems and non-market products (Gulbrandsen 2004; Heynen et al. 2007; Tikina et al. 2010; McDermott 2012).

A second key transformation since the early 1980s has been increased globalisation and liberalisation of trade. While states such as Canada have always been immersed in global trading networks, the shocks and changes in those networks have significantly affected the local economies and social structure of regions of the country throughout our history. Key clauses in many post-1980 international trade agreements shift power from governments to private investors and often contain clauses that grant specific rights to investors allowing them to bypass (and/or be compensated for) the policies of elected governments. International trade agreements, tariff agreements, and the rise of large supranational trading blocs have seen resident natural resource industries in many developed states challenged by the entry of foreign low-cost producers. This trend of global hyper-competitiveness has also resulted in successive crises of profitability within various resource industry sectors. These market factors lead to increased automation, calls to relieve firms of tax and other costs, including the lessening of environmental standards, and demands for increased access to resources so as to boost output, volume-based productivity and, ultimately, competitiveness with producers found in jurisdictions with lower labour costs and reduced regulatory burdens.

The transformation from a Keynesian to a neoliberal framework has had implications for how the key players in environmental, social, and health impact assessment processes engage with the ideas of cumulative impacts, including considerations of health and well-being. To start, governments have seen their regulatory roles and capacity to monitor reduced over time. They have also seen the scope of their ability to manage the economy, and to offer development guidance and incentives, limited by international trade agreements that specifically curtail practices deemed to give national firms an unfair economic advantage (Markey et al. 2012). Although their extant roles may be decreasing, and although there may be a concomitant withdrawal from previously important policy arenas such as regional development, provincial, state, and federal governments still hold the critical policy and fiscal decision-making levers. As will be noted shortly, these have not been devolved to local or regional governments. This creates a mismatch, where the state is less involved, but at the same time continues to hold the critical decision-making authority—all the while with a diminishing level of capacity to act.

The political restructuring aspects of this transformation highlight issues of both government and governance that were introduced in Sect. 1.6 (see also Graham et al. 2003; Jordan et al. 2003). In keeping with the idea of *government* as the "activities undertaken primarily or wholly by state bodies" (Jordan et al. 2003, p. 8), many have argued that the neoliberal turn has increased the burden and posed new challenges for local government in rural and small town places, especially as a result of decentralisation and off-loading by senior levels of government without a concomitant transfer of fiscal resources to local government to deal with those responsibilities (see Aarsæther and Bærenholdt 2001; Odagiri and Jean 2004; Douglas 2005). A concomitant shift towards *governance* demands increased attention to diverse processes of participation and involvement in decision-making, raising issues of power and control that may be dispersed among a variety of state and non-state actors representing a greater diversity of interests and institutions (see Chap. 1; Marsden and Murdoch 1998; Douglas 2005; Pemberton and Goodwin 2010). In smaller places, there can be a very real fragility to governance structures due to the small numbers of available participants and local institutions.

The shift from a top-down government-managed regime for assessing and monitoring resource development towards more of a governance regime has several implications. To start, governance

> implies a re-drawing of the lines of accountability and control, away from centralized state power, to be dispersed amongst a greater diversity of local and extra-local actors and institutions. Second, the participation inherent in governance fosters a sense of ownership, over decisions and, ultimately, resources, that may not have existed under previous top-down regimes. (Markey et al. 2008, p. 411)

Amin and Thrift (1994) refer to the concept of *institutional thickness* to describe the possibilities associated with collaborative governance. Important for our argument about the need for a paradigm shift in how we approach cumulative impact assessment is that a governance (as opposed to government) approach also allows for experimentation with different institutional structures and relationships (Storper 1999; MacLeod 2001; Scott 2004).

Industry and communities have stepped into the regional development breach left by the neoliberal state. Starting with industry, the traditionally risk-averse nature of large corporations that had been supported by society to take risks under a Keynesian framework is now more exposed to market fluctuations under a neoliberal framework. In BC, through the 1950s and 1960s, it was visionary public policy that attracted industrial investments in the form of aluminium smelting and the pulp and paper industry. Today, public policy follows market signals, and industry pursues risk-averse investment decisions and well-established products and associated markets. Proposals to sell LNG to Asia are being driven by firms that market gas, understand their industry, see an economic opportunity and wish to close on that opportunity before the profit margins narrow. Public policy has shifted orientation towards creating conditions that support resource industry success, often after development proposals are already on the table.

This restructuring of roles has also been played out at the local and regional level for communities and interest groups. In BC's province building era of the 1950s and 1960s, the provincial government set in place a vision for large-scale resource development, recruited industrial partners and proceeded with the public investments needed to complement and realise private sector investments (Mitchell 1983; Williston and Keller 1997). Under this model, communities and regions had little voice in decisions or outcomes.

The transition into a neoliberal framework has meant a twofold change for communities. First, on the argument that they should have more responsibility for development planning at the local level, they have been successful (Jackson and Curry 2002). Counter to this increasing responsibility, however, provincial governments have not downloaded additional policy or decision-making authority. The second is that with increasing responsibility, even without the critical public policy levers that the province retains, there has been a notable lack of new fiscal resources to help build the necessary capacity to engage in local and regional development planning. As an example, the municipality of Kitimat, BC, has a local economic development department and an industrial base that experienced significant contraction over the past decade. In short order, several large gas and oil pipeline terminals have recently been proposed. Two factors challenged the fulfillment of a local development vision. First, the scope of the proposed activities very quickly overwhelmed the capacity of the small local government office (even though Kitimat has a very good and a very professional local government). Second, the decision-making authority and public policy levers needed to support Kitimat in adjusting to the scope and scale of this new industrial activity remain with the province. Into early 2014, the province had still not allocated additional services or funding support to assist with local economic development planning and the active community transition that was already underway. The municipality is struggling to accommodate even the most obvious of the cumulative impacts of the preliminary stages of these proposed projects. Among these first obvious impacts have been a rise in poverty and homelessness as housing costs spiralled and low-income renters were evicted, a rise in pressure on the local food bank, the need to create a supplementary food share program, the need to augment social services to address issues around household and personal stress, increased traffic, and increased pressure on health care services and providers from the large influx of temporary workers (Ryser and Halseth 2013). The downloading of responsibility without decision-making authority and without significant new capacity has left many small communities, such as Kitimat, challenged to participate in significant dialogue about cumulative effects and impacts.

Another important consideration is the particular experiences of Aboriginal communities that must attempt to accommodate the cumulative impacts of industrial development that can be counter to strong cultural ties to the land. This issue has relevance to First Nations, Métis, and Inuit peoples in Canada (Parlee et al. 2007; Booth and Skelton 2011; Parlee and Furgal 2012; Richmond 2015) and across diverse Indigenous peoples internationally (Nettleton et al 2007; Heinämäki 2010; Sharapov and Shabayev 2011; Trigger et al. 2014). Addressing such impacts is made more difficult by several hundred years of crippling colonisation that has left Aboriginal communities with little internal or financial capacity to monitor change across their traditional territories or to proactively develop decision-making relationships and development agreements with government and industry (see Sect. 1.7; Vignette 6.6). Issues of concern long voiced by Aboriginal peoples and enunciated in supportive reports and commission (see for example Canada's Royal Commission on Aboriginal Peoples 1996) include the need for a resolution of historic injustices, proper recognition of treaty, rights and title issues, and ongoing and increasing calls for local responsibility (especially in the fields of economic and community development) (Tennant 1990; Fondahl 1998; Harris 2002; Richmond 2015). There has been relatively little assistance, however, for these communities to develop the capacity or policy supports to effectively address vital economic and community development tasks, although Supreme Court decisions continue to highlight that these issues cannot be overlooked (see Boxes 5.1, 5.4; Vignette 6.6). As with non-Aboriginal communities, these challenges and imperatives have only increased following the shift from a Keynesian to neoliberal framework.

In summary, the evolution of environmental, social, and health impact assessment has occurred against a backdrop of social, political, and economic restructuring. This restructuring is complex and multifaceted, but close to its core is the transition from a Keynesian to neoliberal political economy framework. This transition has shifted the roles and capacities of the key actors in the cumulative impact debate. Where once government had the leading and sole responsibility for reviewing projects, monitoring long-term impacts, accommodating the needs of affected communities and maintaining environmental values, now we find that the private sector is being asked to take on more of that role. The outcomes of neoliberal approaches for measuring and addressing cumulative impacts are unclear (Castree 2008). This approach to governance, especially across public lands, however, underscores the need to seek a new paradigm for understanding and managing the cumulative impacts of natural resource development on health, the environment, and communities.

## 7.4 Environmental Assessment Legislation as a Context for Assessing Cumulative Impacts

In Canada, and internationally, EA has emerged as a common and well-tested approach for addressing immediate project impacts—although the extent to and processes by which EA is mandated by legislation vary across jurisdictions. As outlined in Chap. 2, assessments of cumulative impacts tend to rise as a function of project review, mitigation, and ultimately approval, but is not viewed as an effective process (Duinker and Greig 2006). In Canada, the review of proposed projects is typically related to national or provincial (or both) EA legislation. The origins of Canada's EA legislation can be traced to the US precedents of the late 1960s. The US *National Environmental Policy Act* (United States Federal Government 1969) was a top-down legislative response to the apparent impacts resulting from large-project development that was later amended to consider cumulative impacts (see Kenna 2011).

Following the example of the USA, many nations have developed EA legislation that often included a consideration of cumulative impacts (e.g. Canada, China, Egypt, EU, the Netherlands, Hong Kong, Malaysia, Nepal, New Zealand, Russian Federation, Sri Lanka, and Sweden). The establishment and development of the International Association for Impact Assessment in 1980 is one indication of the growing and ongoing attention to these issues (IAIA 1980). Some 40 years of application of traditional project-specific EA by numerous nations has led to a growing and convincing critique of this approach. At the heart of this criticism is a failure to fully consider the cumulative impacts that may extend beyond the reviewable project and to place those impacts within the context of some vision of regional sustainability. Although these criticisms are now widely recognised, there has been little progress towards addressing impacts in their broader, strategic, or cumulative context (see for example Partidário 2012).

Related to the underlying technical and philosophical flaws in project-specific assessment is a trend of further codification and restriction of process (Doelle 2012). The simplification of project review does not match the increasing complexity and urgency of assessing and addressing cumulative impacts, as witnessed across many jurisdictions. Where we should be developing more holistic processes, governments are now restricting scope, and greater scope is what is required to better understand and manage a larger range of environmental, social, and health impacts.

Canada provides a notable example of an evolving legislative context that has seen a clear evolution of process towards a more technical exercise with less scope and more limits on public input (Gibson 2012). The latest version of the *Canadian Environmental Assessment Act*, 2012 (Government of Canada 2012) is meant to be an "updated, modern approach that responds to Canada's current economic and environmental context" (Government of Canada 2014). Although there is still much room for evaluation, it is easy to argue that the *Act* has reduced the comprehensive nature and scope of powers of the process (Doelle 2012; Gibson 2012; Kirchhoff et al. 2013). Relative to the previous version, *Canadian Environmental Assessment Act, 2012* has changed the EA process in a number of ways that will significantly influence the ability of the public or government to better assess and manage cumulative impacts. Specifically the 2012 *Act*:

- Provides less opportunity and shortened timelines for public involvement.
- Limits the range of impacts considered under the *Act* to a precisely defined set of environmental components—there is no direct consideration of social or health impacts beyond Aboriginal communities.
- Focuses application of the Act exclusively on large or controversial projects, greatly reducing the ability of the Act to consider small projects with potentially large cumulative impacts.
- Further enforces a reactive response by the proponent to regulatory requirements, limiting strategic design and mitigative measures that might lessen cumulative impacts.

Further details regarding limitations and challenges of the 2012 *Act* in relation to cumulative environmental, community, and health impacts are elaborated elsewhere (see for example Boxes 5.1 and 5.4 in Chap. 5, and Vignettes 6.6 and 6.8).

Although specific to BC and Canada, Table 7.1 highlights the rapid change in policy and legislation; thus, efforts to develop processes to assess cumulative impacts are occurring in relation to a moving target. Although the *Canadian Environmental Assessment Act* (Government of Canada 2012) requires proponents to consider cumulative impacts, the new legislation does not outline a better process or demand more effective assessment practices. As with the original *Act* (Government of Canada 1992), required CEA remains largely focused on project-specific review with no emphasis on landscape/regional level issues, or the integration across environment, community, and health concerns. Regional studies are formally recognised in the Act, but they are not required and few such studies are conducted. In fact, it can be argued that *Canadian Environmental Assessment Act, 2012* is less effective at addressing cumulative impacts as fewer projects will be assessed and there is a reduction in the range of impacts to be considered (Gibson 2012).

Current EA process and legislation in Canada are focused on the review and mitigation of impacts arising from individual, mostly large, development projects. Work by many academics and practitioners (e.g. Cashmore 2004; Duinker and Greig 2006; Duinker et al. 2012) highlights the range of technical and philosophical limitations of assessing cumulative impacts in the context of an individual development proposal. These critiques underscore key features that need explicit attention in order to improve future practice—reinforcing unmet needs and the imperative for change identified in earlier chapters.

Considering the foundational aspects of CEA, we identified three fundamental limitations of current practice and areas where improvement is much needed. First is the challenge of identifying the scope of area and timescale of effects and impacts. This is especially true of future change because much development is a product of markets and markets are notoriously difficult to predict even for the experts. Second is the procedural paralysis that accompanies assumed or actual deficiencies in information and knowledge. Although we will always be short of information, there are a large number of tools, approaches, and strategies for designing and conducting

**Table 7.1** Evolution of environmental assessment policy, practice, and regulation in Canada and British Columbia

| Year | Policy, practice or regulation | Source |
|---|---|---|
| 1969 | United States *National Environmental Policy Act* | United States Federal Government (1969) |
| 1973 | Canadian Environmental Assessment and Review Process (CEARP) | Noble (2009) |
| 1977 | Berger Inquiry | Berger (1977a, b) |
| 1983 | Formal guidance for practice of environmental assessment: An Ecological Framework for Environmental Impact Assessment | Beanlands and Duinker (1983) |
| 1981 | *BC Environmental Management Act* provides discretionary power to minister to request environmental assessment | Government of BC (1981), Haddock (2010) |
| 1984 | Creation of the Canadian Environmental Assessment Research Council to support EA research | See Hegmann et al. (1999) |
| 1992 | *Canadian Environmental Assessment Act* | Government of Canada (1992) |
| 1994 | National conference focused on a review of the practice of conducting cumulative effect analyses in Canada | See Hegmann et al. (1999) |
| 1994 | *BC Environmental Assessment Act* and formation of the provincial Environmental Assessment Office | Government of BC (1994) |
| 1999 | Formal guidance for assessing cumulative impacts: Cumulative Effects Assessment Practitioners Guide | Hegmann et al. (1999) |
| 1999 | Formal guidance for practice of health impact assessment: A Canadian Health Impact Assessment Guide Volume 2: Decision Making In Environmental Health Impact Assessment | Health Canada (1999, 2004) |
| 2002 | *BC Environmental Assessment Act* (1994) repealed, replaced with revised *BC Environmental Assessment Act* (2002) | Government of BC (2002) |
| 2010 | *BC Environmental Assessment Act* amended as part of *Clean Energy Act* to consider cumulative effects, but this is a discretionary requirement | Government of BC (2010), Haddock (2010) |
| 2012 | *Canadian Environmental Assessment Act, 1992* repealed, replaced with revised *Canadian Environmental Assessment Act, 2012* | Government of Canada (2012) |

CEA that are not being used effectively to influence decision-making and land-use policy (Schneider et al. 2003). Alternatively, there are many situations where we know enough to understand that future development will further degrade water and air quality or result in the loss of endangered species. The rapid development of oil and gas reserves in BC and Alberta is an example of human activities with obvious and significant impacts for human communities and the natural environment. Third, existing approaches and experiences have tended to focus on environmental values, with limited attention to the health or community impacts arising as a consequence of environmental effects, or as direct impacts from development activities.

Even where we might address the foundational limitations of our current approaches for considering cumulative impacts, the overwhelming emphasis on project-specific assessment, as dictated by current legislation, is widely recognised as fundamentally flawed and beyond repair (Duinker and Greig 2006; Jay et al. 2007). Drawing on ideas presented by others and in previous chapters, current EA processes for addressing cumulative impacts are criticised as being:

- Reactive to individual development projects rather than being strategic or proactive.
- Focused across small areas not regions.
- Giving too much emphasis to bottom-up approaches starting with individual projects and proponents as compared to top-down approaches that include all stakeholders across a region.
- Limited in relation to issues of cross-sectoral cooperation and potential efficiencies gained through industry, government, and community collaboration—benefits that would increase predictability of process for communities and proponents, reduce costs for all, and lead to outcomes that better address potential impacts.
- Unable to provide context for regional sustainability including the long-term vision and wants of local interests that will assume the negative (and positive) impacts of development.

An extensive international literature clearly supports these criticisms. In Canada, for example, we can trace initial concerns and efforts for better practices to the early 1990s, only a few years after the *Canadian Environmental Assessment Act, 1992* (Government of Canada 1992), but there has been little progress in better addressing cumulative impacts. Thus, our struggles with EA and more broadly the management of cumulative impacts are not the product of a lack of thought or critical inquiry. We now have 20 years of reflection, but no significant inroads on how to better manage cumulative impacts.

Our failure to make progress is troubling as the rate and scale of resource development, and by consequence environmental pressures are increasing. This is especially concerning, as looking to the future we can expect cumulative impacts to increase rather than decrease (Chap. 2). As introduced in Chap. 2, we do have some experience developing assessment and management frameworks that focus on cross-sectoral issues at regional scales. Referred to as REA or in Canada as RSEA (e.g. CEAMF) they are the product of specific cumulative impact challenges. As a result, they have followed no standard objectives, operational practices, or governance structures. Some have focused on a technical assessment of current and future impacts across sensitive regional areas whereas others involve a fuller spectrum of stakeholders and issues resulting in a more complete framework for both assessing and managing impacts. Despite the range of approaches and objectives, these approaches have had limited success (Johnson 2011). Reasons for failing to manage or reduce impacts include: governance structures that limit the capacity for decision-making; insufficient baseline data, data sharing and monitoring; poor coordination and leadership; and difficulties in matching operational thinking with strategic priorities (Gunn and Noble 2009; Noble et al. 2013).

As highlighted by current efforts to address cumulative impacts, an increasingly important question is who convenes and manages the process? In the past, it was government that spearheaded multi-partner collaboration, involving various levels of government, industry, direct Aboriginal and non-Aboriginal community representation, as well as environmental non-governmental organizations and other interest groups. The CEAMF process for petroleum development in the Athabasca watershed, Alberta, Canada, offers a good example of how a lack of leadership and coordination has limited efforts to address cumulative impacts (Noble et al. 2013). A contrasting approach is an increasing tendency for industry to assume the role of convener and facilitator of these assessment processes. This reflects the wider trend towards neoliberal resource management that has restructured and diminished the role for government.

## 7.5 Converging Challenges and the Integration Imperative

In this chapter, we have framed the dilemma facing us: cumulative impacts need to be understood and managed within the larger context of historical land-use and natural resource policy as well as conceptual and technical challenges. In the sections above, we have contextualised the need for change in relation to the transformative work and findings of the Berger Inquiry. We see the need for a *neo-Berger moment* whereby existing EA and, more broadly, land-use policy is transformed to incorporate factors that have been previously ignored.

This final section highlights common patterns and converging challenges that are emblematic of what is not working with the current processes and why new approaches are required. We describe these converging challenges in relation to the *integration imperative* that frames this book, and explore this imperative with a particular focus on two key features. A first feature of this integration imperative is that it is *relational*, highlighting the need to frame and characterise the relevant components and considerations in relation to each other. This has been exemplified throughout the text, but especially by the Vignettes in Chap. 6, that demonstrate the range of ways that the environment, community, and health dynamics cannot be fully understood in isolation, but rather need to be considered in relation to each other. Second, these considerations should be considered as active and ongoing—an *integrative* challenge—rather than a static or completed act of integration. The need for a dynamic engagement with integration is especially important in relation to the issues of spatial and temporal scale, timing and the past-present-future dynamics introduced in Chap. 1 (see also Box 1.1) and exemplified in later chapters.

It is important to note that these relational and integrative features are neither new nor unique to the challenge of assessing cumulative impacts. Indeed, calls for integration across environment, health, and community concerns have been the focus of boundary-crossing scholarly and policy attention for decades (e.g. Suter 1997; Klein 2000; Bammer 2005; Hallstrom et al. 2015). This relational and integrative emphasis is also consistent with an (eco)systemic approach to understanding cumulative impacts, drawing on several decades of ecosystem approaches to community, health, and sustainability challenges (see for example, Waltner-Toews et al. 2008; Parkes et al. 2010; Webb et al. 2010). Although the challenge of integration is not new, what is important for our purposes is the particular focus on the relational and integrative features in conjunction with the cumulative and regional dynamics of impact assessment.

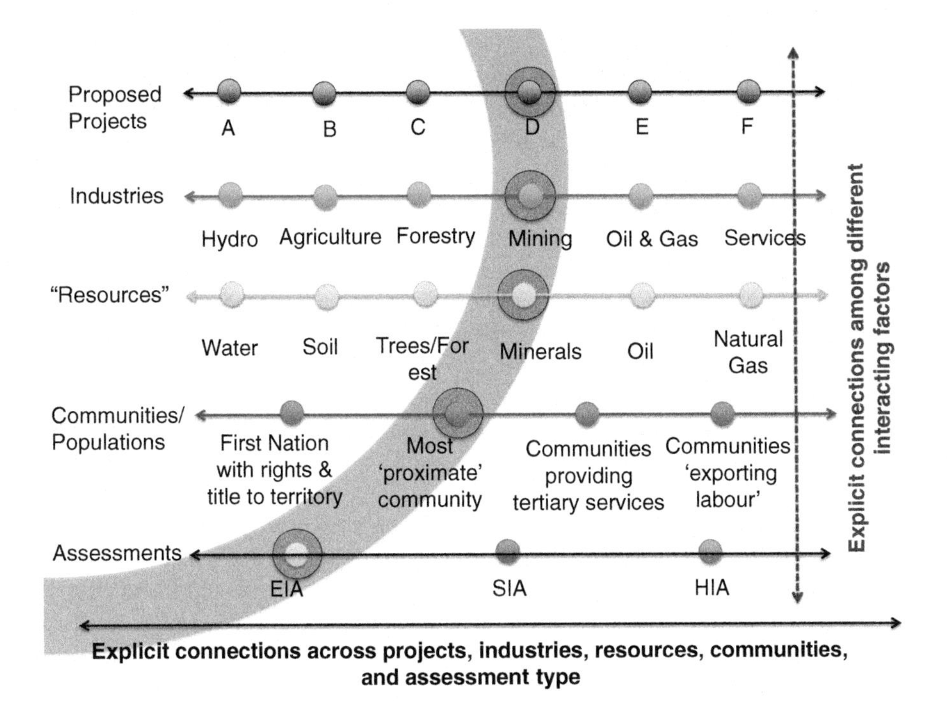


**Fig. 7.1** Moving beyond the limitations of the singular gaze. The shaded arc represents a limited singular gaze, highlighting one project, within one industry, focused on one resource, one community and one kind of impact assessment and excluding things outside of this view. Cumulative thinking begins when we shift our gaze to acknowledge the wider backdrop of change, and bring other interrelated factors into view. This encourages explicit connections across projects, industries, resources, communities, and assessments (*horizontal arrows*), and also connections among these interacting factors (*vertical arrow*)

The imperative of relational and integrative features of cumulative impacts becomes especially apparent when we consider what happens when they are overlooked. The failure to link across and among the different projects and their spatial and temporal context means that we are not only missing *additive* dimensions, but also *emergent* and unexpected factors which are invisible and, therefore, overlooked within the assessment of any one particular project. Figure 7.1 provides a depiction of different considerations, and highlights the way in which a *singular gaze* on any one component of cumulative impacts will result in a failure to recognise additive and emergent dynamics among the other factors.

Working through the considerations depicted in Fig. 7.1 underscores problems that arise when we fail to consider the connections and interactions across projects, industries, resources, communities, or assessments. For example, different government agencies may be charged with the approval and management of particular types of industrial development—an oil and gas proposal may be assessed with a different set of parameters and regulated using a different piece of legislation when compared to a mining or forestry proposal. The resulting lack of coordination and inconsistencies confuses land-use decision-making and provides opportunities for unintentional, but potentially significant cumulative impacts. Each of the horizontal links in Fig. 7.1 underscores the value of shifting from a singular gaze and making explicit *horizontal* connections, such as the integration of EIA, SIA, and HIA within a more comprehensive cumulative impacts framework.

The connections depicted in Fig. 7.1 also highlight ways in which cumulative dynamics can become invisible if we fail to link *vertically* among projects, industries, resource types, communities, and assessment processes. When the connections among these different interacting factors are not made explicit, we tend towards bilateral engagements instead of multilateral relationships among different sectors or different levels of government. This lack of integration prevents recognition and delegation of responsibility for considering the whole *cumulative* picture. The lack of vertical integration has only increased as a result of neoliberal decision-making. Where in the past government assumed the role of steward for public lands, including the coordination of all decision-making, there has been a move to more direct relationships between industry and communities. This *leapfrogging* phenomenon may have some advantages for affected communities through an increase in dialogue and benefit sharing. It may also be beyond the capacity of smaller communities who would then be more vulnerable in the negotiation and legal agreement process. However, ad hoc bilateral relationships between communities and industry will only lessen coordinated decision-making among and even within sectors. This is a movement away from integration with a likely increase in cumulative impacts across regional areas.

Although a focus on individual projects, communities, or resources may seem expedient, they are in effect *shortcuts* that—when taken in isolation and not treated within their larger context—create a built-in failure to acknowledge interrelationships. This then leaves us prone to unintended consequences. Drawing on the depiction in Fig. 7.1, a brief sketch of some of ways in which a *singular gaze* limits and hampers our attention to cumulative impacts include the tendency to focus on:

- Individual projects (e.g. mining project, oil and gas project, forestry clear-cut within one area);
- Individual industries (e.g. forestry, oil and gas, mining, urban development, all with their different legislative contexts and rules);
- Individual resources (e.g. water, soil, trees, each with their different legislative contexts, but which may span projects and industries);
- Individual communities with each navigating their own relationships with other levels of government or directly with industry;
- Individual assessments that fail to explore the cumulative dynamics within and across individual environmental, social, and human impact assessments; and
- Individual/discrete spatial units that are seen separately rather than nested, such as a watershed having a reach, sub-watershed, watershed, and river basin.

In the context of the social–ecological systems within which resource development occurs, a singular gaze on any of the factors above creates a blinkered orientation and tunnel vision (at the extreme, a type of blindness) that has far-reaching consequences in the short, medium, and long terms. What becomes *ecologically* invisible is the emergence of different types of impacts. What becomes *socially* invisible is the need for collective, as compared to individual processes, in order to understand and address these impacts. Some would argue that it is simply not possible to address all the considerations in Fig. 7.1 due to limitations of time, methods, and governance processes. Yet simply by listing or describing each of the considerations, it is clear that none should be overlooked. Rather, seeing these factors together highlights the demand for new ways of thinking and acting in response to cumulative impacts—not only how we gain a more integrative and comprehensive understanding of the rate and scale of changes around us, but also how we can develop adequate responses that reflect this new understanding.

Building on ideas noted in Chap. 1 and earlier in this chapter, the widespread nature and consequences of these dilemmas are being increasingly referred to as *wicked problems* that are characterised by interconnected causes and conflicting interests which are typically not amenable to one simple cure (see Churchman 1967; Brown et al. 2010). Wicked problems are difficult to clearly define; attempted solutions often result in unintended and interdependent consequences; they are dynamic, resulting in no clear target for any one solution; they are socially complex, rarely falling under a single organisational mandate; and they require changes in organisational behaviour often characterised by chronic policy failure (APSC 2007). Importantly, considering these issues as wicked problems is helping to shift a new level of attention to issues that are embedded in society, display complex interdependency that escapes simple definition, are not solvable by addressing manageable sub-problems, and often result in unintended consequences when efforts to solve one aspect of a problem reveal or create other problems (APSC 2007; Brown et al. 2010).

When the characteristics of wicked problems are not adequately acknowledged, there is also the attempt to tweak, refine, or dysfunctionally respond to current challenges. This creates an unnecessary series of false dichotomies and polarisations that compound rather than correct the issue. Acknowledging long-standing recognition of the social and ecological perils of false dichotomies (see Ehrenfeld 1981; Stanley 1995), Table 7.2 provides examples of dichotomies that seek to simplify complex, cumulative realities into a false sense of *either/or* decisions that also tend to ignore the social, historical, and political context that has been emphasised throughout this chapter.

A helpful extension of *both/and* approaches (Table 7.2) is to focus on a continuum, which is especially relevant for the uncertainties involved with cumulative impacts. Recognition of a continuum of certainty is likely to be more helpful than focusing on whether or not there is *enough* data/information or *inadequate* data/information. Issues of spatial and temporal scale are especially obvious examples of the utility of a continuum of certainty (recognising that as spatial scale or timescale increases, certainty may decrease), and serve as an important point of reference when considering a new generation of *cumulative thinking*. The acknowledgement of uncertainty as a characteristic of complex and systemic issues offers an important advance in our ability to understand and address wicked problems, and is well suited to a precautionary and integrative orientation towards ecological, social, and health challenges (Waltner-Toews et al. 2008; Hallstrom et al. 2015).

**Table 7.2** Cumulative thinking beyond false dichotomies: Moving from *either/or* to *both/and* approaches

| False dichotomy (*either/or*) | Examples of *both/and* approaches associated with cumulative thinking |
|---|---|
| Proactive versus reactive | Allowing planning processes to guide reactive development decisions |
| Regional versus national interests | Recognising a nested set of development wants and impacts that includes a range of potentially conflicting interests from the regional setting of a development, through to national interests |
| Efficiency versus Integration | Integration requires appropriate time and inclusivity to fully consider a land-use decision, even if that compromises efficiency as defined by a rapid decision-making process |
| Demand driven versus data (evidence) driven | The demands of proponents for rapid decisions must not compromise the need to collect and fully consider both the necessary information and knowledge for finding the balance between the positive and negative impacts of development |

## 7.6 Conclusions

This chapter commenced with recognition that concerns about the effects and impacts of resource development projects on landscapes and communities are not new. Informed by years of experience with environmental, social, and health impact assessment, concern about our inability to better understand and manage for the cumulative impacts of resource development is also not new. Today, we clearly recognise that a more comprehensive understanding of cumulative impacts will demand attention to the dynamic interactions of environmental, social, and health impacts. It will also demand a paradigmatic shift in policy and practice around cumulative impact assessment.

In this chapter, we have argued that there has been considerable development of our tools and understandings of cumulative impact assessment over the past 40 years. We have also argued that while there has been progress, there have also been increasing numbers of cracks apparent in our existing processes and frameworks. These include concerns about limited spatial and temporal scales, the inadequacies of processes that have become increasingly elaborate dances choreographed by teams of competing lawyers, with a focus on large mega-projects to the exclusion of smaller projects and activities, and more recent changes to speed up or streamline processes by excluding voices. Building from the discussion in Chaps. 1–6, we have underscored that our current approaches to impact assessment (spanning environmental, social, and health dynamics) are all found to be wanting.

As a result, we do not argue for tinkering with a broken process or framework. We argue instead for dramatic change that will demand innovation and integration in how we approach cumulative impact assessment around natural resource development. Although some may be stuck in their current regulatory and approval processes, we opened this chapter by reminding us all that our current systems and processes emerged out of a past set of challenges—and that now is the time for another such paradigmatic shift. Past experience tells us that dramatic and fundamental change is possible.

Not only is such change possible, but this chapter also reminds us that much of the underlying and fundamental building blocks of the older frameworks have been changed through social, political, and economic restructuring. Although some issues remain (the desire of the public to be involved, and the need for clarity of process and transparency in information) the world is also a quite different place today.

Processes born of the 1970s are no longer a good fit for the expectations or the capacities of contemporary governments, industries, or communities. The governance regime around resource development and impact assessment and monitoring is different, and the need for cumulative thinking and approaches is different. By detailing the limitations of current approaches, 40 years of research and experience tells us that the wicked problem of cumulative impact assessment needs a revolutionary shift towards better understandings and solutions. From our critique we know that a revolutionary approach will of necessity be *integrative* and *relational*—highlighting that environment, community, and health dynamics cannot be fully understood in isolation, but must be considered in relation to each other. From this critique, and the integration imperative it describes, Chap. 8 details the elements of a more revolutionary approach.

## References


Aarsæther, N., and J.O. Bærenholdt. 2001. Understanding local dynamics and governance in northern regions. In *Transforming the local*, ed. J.O. Bærenholdt and N. Aarsæther, 15–42. Copenhagen: Nord.

Amin, A., and N. Thrift. 1994. *Globalization, institutions and regional development in Europe*. Oxford: Oxford University Press.

APSC (Australian Public Services Commission). 2007. *Tackling wicked problems: A public policy perspective*. Barton: Australian Public Services Commission, Australian Government.

Bammer, G. 2005. Integration and implementation sciences: Building a new specialization. *Ecology and Society* 10(2): 6.

BC Utilities Commission. 1994. *Kemano completion project review: Summary report*. Vancouver: BC Utilities Commission. http://www.bcuc.com/Documents/Proceedings/2007/DOC_17296_C21-4_Kemano-Project-Summary-Report.pdf. Accessed 5 Nov 2014.

Beanlands, G.E., and P.N. Duinker. 1983. *An ecological framework for environmental impact assessment in Canada*. Halifax: Institute for Resource and Environmental Studies, Dalhousie University.

Berger, T.R. 1977. *Northern frontier, northern homeland: The report of the Mackenzie valley pipeline inquiry—volume one. Social, economic, and environmental impact*. Ottawa: Minister of Supply and Services Canada.

Berkes, F., J. Colding, and C. Folke. 2003. *Navigating social–-ecological systems: Building resilience for complexity and change*. New York: Cambridge University Press.

Booth, A.L., and N.W. Skelton. 2011. Industry and government perspectives on First Nations' participation in the British Columbia environmental process. *Environmental Assessment Impact Review* 31: 216–225.

Brown, V.A., J.A. Harris, and J.Y. Russell. 2010. *Tackling wicked problems: Through the transdisciplinary imagination*. London: Earthscan.

Burdge, R.J., and F. Vanclay. 1995. Social impact assessment. In *Environmental and social impact assessment*, ed. F. Vanclay and D.A. Bronstein, 31–66. New York: John Wiley and Sons.

Cashmore, M. 2004. The role of science in environmental impact assessment: Process and procedure versus purpose in the development of theory. *Environmental Impact Assessment Review* 24: 403–426.

Castree, N. 2008. Neoliberalising nature: The logics of deregulation and reregulation. *Environment and Planning* 40: 131–152.

Churchman, C.W. 1967. Wicked problems (Guest Editorial). *Management Science* 14: B141–B142.

Doelle, M. 2012. CEAA 2012: The end of federal EA as we know it? *Journal of Environmental Law and Practice* 24: 1–17.

Douglas, D. 2005. The restructuring of local government in rural regions: A rural development perspective. *Journal of Rural Studies* 21: 231–246.

Duinker, P.N., and L.A. Greig. 2006. The impotence of cumulative effects assessment in Canada: Ailments and ideas for redeployment. *Environmental Management* 37: 153–161.

Duinker, P.N., E.L. Burbidge, S.R. Boardley, and L.A. Greig. 2012. Scientific dimensions of cumulative effects assessment: Toward improvements in guidance for practice. *Environmental Review* 21: 40–52.

Ehrenfeld, D.W. 1981. *The arrogance of humanism*. New York: Oxford University Press.

Fondahl, G. 1998. *Gaining ground? Evenkis, land and reform in southeastern Siberia*. Wilton: Allyn and Bacon.

Gibson, R.B. 2012. In full retreat: The Canadian government's new environmental assessment law undoes decades of progress. *Impact Assessment and Project Appraisal* 30: 179–188.

Government of BC. 1981. *Environmental Management Act, 1981* (B.C. Reg. 330/81, O.C 1752/81). http://www.bclaws.ca/Recon/document/ID/freeside/23_330_81. Accessed 12 Jan 2015.

———. 1994. Environmental Assessment Act, 1994. (R.S.B.C 1996. C. 119). Repealed by the Environmental Assessment Act, SBC2002, c. 43, s. 58, effective December 30, 2002 (B.C. Reg 370/2002). http://www.bclaws.ca/civix/document/id/complete/statreg/96119rep_01. Accessed 12 Dec 2014.

———. 2002. Environmental Assessment Act, 2002 (S.B.C. 2002, C. 43). http://www.bclaws.ca/civix/document/id/complete/statreg/02043_01. Accessed 10 Dec 2014.

———. 2010. Clean Energy Act, 2010 (S.B.C. 2010, C. 22). http://www.bclaws.ca/civix/document/id/complete/statreg/10022_01. Accessed 12 Dec 2014.

Government of Canada. 1992. Canadian Environmental Assessment Act, 1992 ((S.C. 1992, c. 37). http://laws-lois.justice.gc.ca/eng/acts/c-15.2. Accessed 12 Dec 2014.

———. 2012. Canadian Environmental Assessment Act, 2012 (S.C. 2012, C. 19, S. 52). http://laws-lois.justice.gc.ca/eng/acts/c-15.21/page-1.html. Accessed 12 Dec 2014.

———. 2014. Overview: Canadian Environmental Assessment Act, 2012 Canadian Environmental Assessment Agency. Date Modified 29 July 2014. https://www.ceaa-acee.gc.ca/default.asp?lang=en&n=16254939-1. Accessed 10 Apr 2015.

Graham, J., B. Amos, and T. Plumptre. 2003. *Principles for good governance in the 21st Century. Policy brief no. 15—August 2003*. Ottawa: Institute on Governance. http://iog.ca/wp-content/uploads/2012/12/2003_August_policybrief15.pdf. Accessed 21 Nov 2014.

Gulbrandsen, L.H. 2004. Overlapping public and private governance: Can forest certification fill the gaps in the global forest regime. *Global Environmental Politics* 4: 75–99.

Gunn, J.H., and B.F. Noble. 2009. Integrating cumulative effects in regional strategic environmental assessment frameworks: Lessons from practice. *Journal of Environmental Assessment Policy and Management* 11: 267–290.

Haddock, M. 2010. *Environmental assessment in British Columbia*. Victoria: Environmental Law Centre, University of Victoria.

Hallstrom, L.K., M.W. Parkes, and N. Guelstorf. 2015. Convergence and diversity: Integrating encounters with health, ecological and social concerns. In *Ecosystems, society and health: Pathways through diversity, convergence and integration*, ed. L.K. Hallstrom, N. Guelstorf, and M.W. Parkes. Montreal: McGill-Queens University Press.

Harris, R.C. 2002. *Making native space: Colonialism, resistance, and reserves in British Columbia*. Vancouver: UBC Press.
Harvey, D. 2005. *A brief history of Neoliberalism*. Oxford: Oxford University Press.
Health Canada. 1999. *A Canadian health impact assessment guide volume 1: The basics*. Ottawa: Health Canada: Santé Canada. http://publications.gc.ca/collections/Collection/H46-2-99-235E-1.pdf. Accessed 5 Nov 2014.
———. 2004. *A Canadian health impact assessment guide volume 2: Approaches and decision-making*. Ottawa: Health Canada: Santé Canada. Archived 24 June 2013. http://publications.gc.ca/collections/Collection/H46-2-04-361E.pdf. Accessed 5 Nov 2014.
Hegmann, G., C. Cocklin, R. Creasey, S. Dupuis, A. Kennedy, L. Kingsley, W. Ross, H. Spaling, and D. Stalker. 1999. *Cumulative effects assessment practitioners guide*. Hull: AXYS Environmental Consulting, and the CEA Working Group for the Canadian Environmental Assessment Agency.
Heinämäki, L. 2010. *The right to be a part of nature: Indigenous peoples and the environment*. Rovaniemi: Lapland University Press.
Heynen, N., J. McCarthy, S. Prudham, and P. Robbins (eds.). 2007. *Neoliberal environments: False promises and unnatural consequences*. London: Routledge Press.
Higgins, V., and S. Lockie. 2002. Re-discovering the social: Neo-liberalism and hybrid practices of governing in rural natural resource management. *Journal of Rural Studies* 18: 419–428.
IAIA (International Association for Impact Assessment). 1980. About IAIA. http://www.iaia.org. Accessed 12 Dec 2014.
Jackson, T., and J. Curry. 2002. Regional development and land use planning in rural British Columbia: Peace in the woods? *Regional Studies* 36: 439–443.
Jay, S., C. Jones, P. Slinn, and C. Wood. 2007. Environmental impact assessment: Retrospect and prospect. *Environmental Impact Assessment Review* 27: 287–289.
Johnson, C.J. 2011. Regulating and planning for cumulative effects: The Canadian experience. In *Cumulative effects in wildlife management: Impact mitigation*, ed. P.R. Krausman and L.K. Harris, 29–46. Boca Raton: CRC Press.
Jordan, A., R. Wurzel, and A. Zito. 2003. New instruments of environmental governance. *Environmental Politics* 12(3): 1–24.
Kenna, M. 2011. The NEPA process: What the law says. In *Cumulative effects in wildlife management: Impact mitigation*, ed. P.R. Krausman and L.K. Harris, 17–27. Boca Raton: CRC Press.
Kinnear, S., Z. Kabir, J. Mann, and L. Bricknell. 2013. The need to measure and manage the cumulative impacts of resource development on public health: An Australian perspective. In *Current topics in public health*, ed. A. Rodriguez-Morales, 125–144. Rijeka: InTech Publishers.
Kirchhoff, D., H.L. Gardner, and L.J.S. Tsuji. 2013. The Canadian Environmental Assessment Act, 2012 and associated policy: Implications for Aboriginal people. *The International Indigenous Policy Journal* 4(3): 14–16.
Klein, J. 2000. Integration, evaluation, and disciplinarity. In *Transdisciplinarity: Recreating integrated knowledge*, ed. M.A. Somerville and D. Rapport, 49–59. Oxford: EOLSS Publishers.
MacLeod, G. 2001. New regionalism reconsidered: Globalization and the remaking of political economic space. *International Journal of Urban and Regional Research* 25: 804–829.
Markey, S., G. Halseth, and D. Manson. 2008. Challenging the inevitability of rural decline: Advancing the policy of place in Northern British Columbia. *Journal of Rural Studies* 24: 409–421.
———. 2012b. *Investing in place: Economic renewal in Northern British Columbia*. Vancouver: UBC Press.
Marsden, T., and J. Murdoch. 1998. Editorial: The shifting nature of rural governance and community participation. *Journal of Rural Studies* 14: 1–4.
McDermott, C.L. 2012. Trust, legitimacy, power in forest certification: A case study of the FSC in British Columbia. *Geoforum* 43: 634–644.
Mitchell, D. 1983. *WAC Bennett and the rise of British Columbia*. Vancouver: Douglas and MacIntyre.
Nettleton, C., C. Stephens, F. Bristow, S. Claro, T. Hart, C. McCausland, and I. Miljlof. 2007. Utz Wachil: Findings from an international study of Indigenous perspectives on health and environment. *EcoHealth* 4: 461–471.

Noble, B.F. 2009. Promise and dismay: The state of strategic environmental assessment systems and practices in Canada. *Environmental Impact Assessment Review* 29: 66–75.

———. 2010. *Introduction to environmental impact assessment: A guide to principles and practice*, 2nd ed. Don Mills: Oxford University Press.

Noble, B.F., J.S. Skwaruk, and R.J. Patrick. 2013. Toward cumulative effects assessment and management in the Athabasca watershed, Alberta, Canada. *The Canadian Geographer* 58: 315–328.

Odagiri, T., and B. Jean. 2004. The roles of local governments for revitalization of rural areas in Japan and Canada. In *Building for success: Exploration of rural community and rural development*, eds. G. Halseth, and R. Halseth, 363–392. Brandon: Rural Development Institute, Brandon University.

Parkes, M.W., K.E. Morrison, M.J. Bunch, L.K. Hallstrom, R.C. Neudoerffer, H.D. Venema, and D. Waltner-Toews. 2010. Towards integrated governance for water, health and social–ecological systems: The watershed governance prism. *Global Environmental Change* 20: 693–704.

Parlee, B., and C. Furgal, 2012. Well-being and environmental change in the arctic: A synthesis of selected research from Canada's International Polar Year program. *Climatic Change*. 115: 13–34.

Parlee, B., J. O'Neil, and Lutsel K'e. Dene First Nation. 2007. The Dene way of life: Perspectives on health from Canada's north. *Journal of Canadian Studies* 41: 112–133.

Parlee, B.L., K. Geertsma, and A. Willier. 2012. Social–ecological thresholds in a changing boreal landscape: Insights from Cree knowledge of the Lesser Slave Lake Region of Alberta, Canada. *Ecology and Society* 17(2): 20.

Partidário, M.R. 2012. *Strategic environmental assessment better practice guide: Methodological guidance for strategic thinking in SEA*. Lisbon: Portuguese Environment Agency and Redes Energéticas Nacionais.

Peck, J., and A. Tickell. 2002. Neoliberalizing space. *Antipode* 34: 380–404.

Pemberton, S., and M. Goodwin. 2010. Rethinking the changing structures of rural local government—State power, rural politics and local political strategies? *Journal of Rural Studies* 26: 272–283.

Peterson, E.B., Y.H. Chan, N.M. Peterson, G.A. Constable, R.B. Caton, C.S. Davis, R.R. Wallace, and G.A. Yarronton. 1987. *Cumulative effects assessment in Canada: An agenda for action and research*. Hull: Canadian Environmental Assessment Research Council.

Richmond, C. 2015. The relatedness of people, land and health: Stories from Anishinabe Elders. Chapter 3. In: *Determinants of Indigenous Peoples' Health in Canada: Beyond the Social*, eds. M. Greenwood, S. de Leeuw, N. Lindsay, and C. Reading. Toronto: Canadian Scholars Press.

Royal Commission on Aboriginal Peoples. 1996. *Report of the Royal Commission on Aboriginal peoples, Volume 1—Looking Forward, Looking Back*. Ottawa: The Royal Commission on Aboriginal Peoples.

Ryser, L., and G. Halseth. 2013. *Tracking the Social and Economic Transformation Process in Kitimat, BC: Interim Summary Report*. Prince George: Community Development Institute, University of Northern British Columbia.

Sabin, P. 1995. Voices from the hydrocarbon frontier: Canada's Mackenzie Valley Pipeline Inquiry, 1974–1977. *Environmental History Review* 18: 17–48.

Scarpa, S. 2013. New geographically differentiated configurations of social risks: Labour market policy developments in Sweden and Finland. In *Changing social risks and social policy responses in the Nordic welfare states*, ed. I. Harsløf and R. Ulmestig, 220–245. New York: Palgrave Macmillan.

Schneider, R.R., J.B. Stelfox, S. Boutin, and S. Wasel. 2003. Managing the cumulative impacts of land uses in the Western Canadian Sedimentary Basin: A modelling approach. *Conservation Ecology* 7(1): 8.

Schultz, C.A. 2012. History of the cumulative effects analysis requirement under NEPA and its interpretation in U.S. Forest Service. *Journal of Environmental Law and Litigation* 27: 125–190.

Scott, M. 2004. Building institutional capacity in rural Northern Ireland: The role of partnership governance in the LEADER II programme. *Journal of Rural Studies* 20: 49–59.

Sharapov, V., and Y. Shabayev. 2011. The Izhma Komi and the Pomor: Two models of cultural transformation. *Journal of Ethnology and Folkloristics* 5: 97–122.

Sheppard, E. 2009. Political economy. In *The dictionary of human geography*, ed. D. Gregory, R. Johnston, G. Pratt, M. Watts, and S. Whatmore, 547–549. Oxford: Wiley-Blackwell.
Shortall, S., and M. Warner. 2010. Social inclusion or market competitiveness? A comparison of rural development policies in the European Union and the United States. *Social Policy Administration* 44: 575–597.
Stanley, T.R. 1995. Ecosystem management and the arrogance of humanism. *Conservation Biology* 9: 255–262.
Storper, M. 1999. The resurgence of regional economics: Ten years later. In *The new industrial geography: Regions, regulation and institutions*, ed. T.J. Barnes and M.S. Gertler, 25–53. Routledge: New York.
Sullivan, L., L. Ryser, and G. Halseth. 2014. Recognizing change, recognizing rural: The new rural economy and towards a new model of rural services. *Journal of Rural and Community Development* 9: 219–245.
Suter, G.W. 1997. Integration of human health and ecological risk assessment. *Environmental Health Perspectives* 105: 1282–1283.
Tennant, P. 1990. *Aboriginal peoples and politics: The Indian land question in British Columbia, 1849–1989*. Vancouver: UBC Press.
Tikina, A.V., J.L. Innes, R.L. Trosper, and B.C. Larson. 2010. Aboriginal peoples and forest certification: A review of the Canadian situation. *Ecology and Society* 15: 1–19.
Tonts, M., and F. Haslam-McKenzie. 2005. Neoliberalism and changing regional policy in Australia. *International Planning Studies* 10: 183–200.
Tonts, M., and R. Jones. 1997. From state paternalism to neoliberalism in Australian rural policy: Perspectives from the Western Australian Wheatbelt. *Space and Polity* 1: 171–190.
Trigger, D., J. Keenan, K. de Rijke, and W. Rifkin. 2014. Aboriginal engagement and agreement-making with a rapidly developing resource industry: Coal seam gas development in Australia. *The Extractive Industries and Society* 1: 176–188.
United States Federal Government. 1969. National Environmental Policy Act of 1969. [42 U.S.C. §4321 et seq. (1969)]. http://www2.epa.gov/laws-regulations/summary-national-environmental-policy-act. Accessed 12 Dec 2014.
Vanclay, F., and A.M. Esteves (eds.). 2011. *New directions in social impact assessment: Conceptual and methodological advances*. Cheltenham: Edward Elgar Publishing.
Waltner-Toews, D., J.J. Kay, and N.-M.E. Lister. 2008. *The ecosystem approach: Complexity, uncertainty, and managing for sustainability*. New York: Columbia University Press.
Webb, J., D. Mergler, M.W. Parkes, J. Saint-Charles, J. Spiegel, D. Waltner-Toews, A. Yassi, and R.F. Woollard. 2010. Tools for thoughtful action: The role of ecosystem approaches to health in enhancing public health. *Canadian Journal of Public Health* 101: 439–441.
Wernham, A. 2011. Health impact assessments are needed in decision making about environmental and land-use policy. *Health Affairs* 30: 847–956.
Williston, E., and B. Keller. 1997. *Forests, power, and policy: The legacy of ray Williston*. Prince George: Caitlin Press.
Yukon Government. 2002. *Yukon environmental and socio-economic assessment act*. Whitehorse: Yukon Minister of Justice.

# Chapter 8
# A Revolution in Strategy, Not Evolution of Practice: Towards an Integrative Regional Cumulative Impacts Framework

**Chris J. Johnson, Michael P. Gillingham, Greg R. Halseth, and Margot W. Parkes**

## 8.1 Introduction

Ours is not the first call to reform or rethink the way that we conceptualise and address cumulative impacts. Indeed, this subject has received much attention from both scholars and practitioners with near universal criticism of how we measure, assess, and regulate cumulative impacts (e.g. Cocklin 1993; Ross 1998; Kennett 2002; Noble 2009; Duinker et al. 2012). Duinker and Greig (2006) offer the most pointed criticism, arguing that we need to move beyond evolutionary reforms to revolutionising how we undertake CEA, regulation, and planning.

Progress has been made in Canada to rethink cumulative impacts. Regional approaches have taken numerous forms (see Chaps. 2 and 3), but there is still much room for improvement. In particular, sustainability has been invoked as a guiding principle for documenting and addressing cumulative impacts (White and Noble 2013), but the overwhelming emphasis in natural resource policy and practice has been how the cumulative impacts of industrial development influence only the environment. There has been little focus on the other elements of sustainability, our economy, and the social well-being and health of human populations. We argue that new directions forward must include a formal recognition of the health of ecosystems in combination with the needs of communities. As one reads the vignettes and case studies, it becomes apparent that cumulative impacts involve more than biodiversity and ecosystem processes, but also human landscapes and importantly the interrelationships between us and the natural world.

As introduced in Chap. 1, our primary goal is to increase the fullness and ultimately the complexity of the discussion of what we consider as cumulative impacts and how those impacts influence our lives and our landscapes. We also move beyond

C.J. Johnson (✉) • M.P. Gillingham • G.R. Halseth • M.W. Parkes
The University of Northern British Columbia, Prince George, BC, Canada V2N 4Z9
e-mail: Chris.Johnson@unbc.ca; Michael.Gillingham@unbc.ca; Greg.Halseth@unbc.ca; Margot.Parkes@unbc.ca


M.P. Gillingham et al. (eds.), *The Integration Imperative*,
DOI 10.1007/978-3-319-22123-6_8

just understanding, proposing solutions to real problems confronting concerned citizens, professionals, communities, industry, and governments. These solutions require a more comprehensive consideration of effects as they might impact the environment, communities, and human health and the associated interactions among impacts (Cocklin 1993; Waltner-Toews 2004). Beyond a full, but disaggregated consideration, we believe that a more integrative approach, which recognises the interaction of impacts in all of their forms and takes into account regional dynamics, is a critical next step in providing revolutionary solutions to the challenge of cumulative impacts.

We do not claim that it is easy or simple to assess or integrate impacts that are intertwined within our environments and communities. We take some inspiration and guidance, however, from the work of scholars and practitioners that have come to this problem before us. This final chapter, therefore, commences with a synthesis of key points that have emerged from earlier chapters and supporting literatures. We build on those past criticisms and recommendations and present a general framework for an integrative and regional approach to the assessment and management of cumulative impacts. We propose six principles and five elements that provide the structure for conceptualising an integrative regional cumulative impacts framework that can be adapted to unique regional circumstances. Even failing full construction and implementation of such a framework, we feel that these principles and elements are the starting point for dialogue on how to better address the broad suite of cumulative impacts that are occurring at ever greater rates across developing landscapes.

## 8.2 Addressing Cumulative Impacts Today

Through the production of this book and the contemplation of cumulative impacts, we have come to support the notion of revolution in cumulative impacts thinking and practice. We recognise, however, that our call for revolution echoes the voices of others that have spent some 20 years confronting the challenges inherent to the assessment and management of cumulative impacts. Informed by this rich body of past work, we reiterate those criticisms and add to the diverse and complex solution set (e.g. CCME 2009; Partidário 2012). Also, we draw on several decades of research and practice focused on more integrative approaches to interrelated environmental, community, and health issues (e.g. Rapport et al. 1998; McMichael et al. 1999; Waltner-Toews et al. 2008; Charron 2012). This includes arguments that there is an inherent synergy between local and regional actions and activities involving environments, communities, and economies that demands ongoing awareness at both the local and the regional scale (Pierce 1992; Savoie 1997; Barnes and Gertler 1999; Polèse 1999; Porter 2004; Markey et al. 2012). Although the region is a broader and more appropriate scale at which to understand and evaluate impacts, it also remains a reasonable scale for working towards solutions. We do acknowledge that regional solutions are made difficult by new understandings of governance and the need for a more appropriate balance of power, authority, and representation.

Consistent with that past work, our recommendations are not focused on changes to project-specific regulation (e.g. *Canadian Environmental Assessment Act*; Government of Canada 2012), but larger foundational elements of integrative, regional processes that are applicable across jurisdictional boundaries and apply to any suite of potential effects and impacts. Below, we itemise key points that are recognised hallmarks of broad-scale solutions for addressing cumulative impacts, and that provide overarching points of reference prior to delineating the six principles and five elements of our proposed integrative regional cumulative impacts framework.

#### *Regional Sustainability Should Set Objectives for Cumulative Impacts*

We must consider cumulative effects and impacts within the broader context of environmental, economic, and social factors, and their influence on community health and well-being. The concept of sustainability, although still a subject of debate, has broadly defined principles associated with ecosystems and human well-being that can guide such considerations (Brundtland and World Commission on Environment and Development 1987; Mebratu 1998; Waltner-Toews 2004; McMichael 2006). We argue that the many challenges associated with identifying and managing cumulative impacts underscore the need to recognise the concept of sustainability at a regional scale. In many cases, however, this will demand that citizens and decision makers accept the trade-offs among competing values related to the economy, environment, community, and human health.

#### *Cumulative Impacts Must Be Assessed at Large Scales*

A regional perspective will allow a full accounting of impacts for broadly distributed values. Thus, we should measure and address cumulative impacts over large spatial and temporal scales that are better suited to fully recognise regional dynamics. Strategic thinking and planning is necessary to ensure that a region's vision for sustainability is achieved and that many incremental small changes do not result in unanticipated and unwanted cumulative impacts.

#### *Adopt a Tiered Decision-Making Approach for Managing Cumulative Impacts*

Decision making must occur within a tiered framework that includes project-specific and regional assessments as well as planning approaches to identify and achieve a vision for sustainability that eclipses the impacts of any one development project (Harriman and Noble 2008). In isolation, project-specific EA is ineffective, but will still play an important role for large industrial development projects when placed within the context of regional assessments of larger-scale impacts.

#### *Limits to Anthropogenic Landscape Change May Be Necessary*

We may need to use biophysical, cultural, or economic limits to regulate cumulative impacts at a level that is acceptable within the bounds of regional sustainability. It is important to recognise, however, that ecological thresholds are not regulatory limits. Ecological thresholds are difficult to identify and generalise across systems. Even where identified, strict implementation can be confounded by economic considerations that trump ecological values. *Regulatory limits* recognise thresholds in ecosystem dynamics, but are based on transparent trade-offs in values and are a more realistic expectation for complex social–ecological systems (Johnson 2013).

#### *Cumulative Impacts Are Place Specific*

Large-scale approaches and frameworks for assessing cumulative impacts will need to be structured to meet the unique social and ecological characteristics and challenges facing each region (Gunn and Noble 2009a). There is no single model or formula for addressing cumulative impacts at regional scales.

#### *Effective and Empowered Decision-Making Process Is Important*

A formal and transparent governance structure is necessary to coordinate REA and planning, including information collation, planning structure and implementation, and monitoring. Such processes would need to be cross-sectoral and inclusive of communities, industrial players, and other interests with a stake or perspective on regional development (Noble et al. 2013). For credibility and to be effective, such processes must have some formal capacity to make decisions, not just provide advice to higher-levels of government.

#### *Ensure that Integrative Regional Cumulative Impacts Frameworks Are Adaptive*

We have a limited ability to predict economic trends and environmental change. In addition, the nonlinear and synergistic dynamics inherent to cumulative impacts are often unknown or uncertain. Thus, monitoring and adaptation will be key elements for addressing cumulative impacts at broad-scales and over long time periods (Burton et al. 2014). Effective frameworks must also be adaptive to dynamic decision-making processes including changes in legislation, government policy, and evolving community values (White and Noble 2013).

Many of these key points have already been incorporated into what is generically termed REA or in Canada RSEA (Chap. 2). Such an approach is "designed to systematically assess the potential environmental effects including cumulative effects, of alternative strategic initiatives, policies, plans, or programmes for a particular region" (CCME 2009, p. 6). Regional strategic environmental assessment has a number of objectives including increasing the effectiveness of project-specific assessments and identifying strategic directions and objectives for regional areas, relative to potential future cumulative impacts. Gunn and Noble (2009a) clearly differentiate strategic from project-specific EA. The latter is focused on short- or medium-term time horizons, applies to single or closely related projects, and emphasises project impacts and mitigation as opposed to regional planning and sustainability.

In Canada, the philosophy of RSEA has been put into practice in the form of CEAMFs (see Chap. 2). Although there have been some successes, and we are certain to witness future tests of those ideas (e.g. Chetkiewicz and Lintner 2014), there have also been failures and missed opportunity for truly addressing cumulative impacts at regional scales (Johnson 2011). Thus, there is still much need for practice and innovation in the development of regional approaches that more effectively identify and address the cumulative environmental, community, and health impacts of industrial development and landscape change.

## 8.3 Cumulative Impacts Thinking: Building a New Conceptual Framework

Informed by critiques and converging themes arising from the past, we propose *cumulative impacts thinking* as a necessary foundation for the six principles and five elements of the integrative regional cumulative impacts framework that we present. We argue that cumulative thinking provides a necessary starting point for more fully addressing cumulative impacts as they occur across the environment, communities, and human health. These ideas are meant to provide the structure for conceptualising an integrative framework that can be adapted to unique regional circumstances. We purposefully use the term framework to represent a starting point and the necessary supporting architecture for a solution, not a complete process or definitive endpoint.

The principles and elements are premised on the work of others and the experiences of our colleagues, as presented in this book (Chap. 6). Thus, we have some precedent and evidence for this guidance; our proposal is not suggesting radical departure from contemporary lessons or ideas that have been tested in the cumulative impacts arena. Also, we must emphasise that these recommendations are not conclusive. Rather, these principles and elements can be seen as likely prerequisites that—informed by regional circumstances and context—encourage us to move beyond project-specific consideration of cumulative impacts to a more holistic process. We would expect to see refinement and the addition of new elements, but also a full consideration of each of the six principles. Before presenting the principles and elements necessary for constructing a framework, we discuss the philosophical approach of cumulative impacts thinking, or more simply *cumulative thinking* where one recognises and addresses integration, the truly cumulative nature of the impacts, and the range of scales over which those impacts might occur.

### *8.3.1 Cumulative Thinking: The Integration Imperative*

At the core, this book contributes to understanding and addressing the cumulative impacts problem by encouraging a fundamental shift in perspective: a shift from considering the environment as the purely biophysical, to the environment as inclusive of ecological, economic, community, cultural, and health dynamics (Cocklin

1993). We argue that a central building block for the construction of a meaningful and effective cumulative impacts framework is the formal recognition and ultimately the integration of cumulative environmental, community, *and* health impacts. At present, EA legislation drives the formal evaluation of cumulative impacts with an overwhelming emphasis on changes to the biophysical environment. Health and community are typically considered as downstream impacts resulting from environmental change (Chaps. 2 and 5).

The integrative approach that we propose is not a set of interlinked, but separate assessments for each of the three *sectors* (e.g. Esteves et al. 2012; Harris-Roxas et al. 2012). We believe strongly that we must revolutionise how we conceptualise this problem, not apply tested and tired reductionist methods. Indeed, integrative efforts are consistent with the foundational ideas of sustainability—as economic and social development within the context of environmental protection and function (Mebratu 1998). The precise technical language and process for implementing and assessing integration in the context of cumulative impacts have, however, been inadequately developed and demand the cumulative thinking proposed here.

A starting point for cumulative thinking is a commitment to conceptualising the full range of spatial and temporal scales as well as impacts that occur as a result of human-caused changes to coupled social–ecological systems. A cumulative thinking perspective demands recognition that impacts are not just environmental, they are not just large development projects, and they are not all easily identified and quantified. Furthermore, past experience tells us to expect interactions. Industrial activities and resulting environmental change are motivated by economic development. The generation of wealth for communities and governments will have consequences for the environment with direct or indirect affects on human well-being and health. Considering the complete scope of cumulative impacts is inherently complex, but this should not be a reason for reductionism or a failure to integrate impacts.

The title of the book speaks directly to the *integration imperative*: integration of impacts is essential to both understanding the problem and developing solutions. This is not to claim that the concept and benefit of integration is novel. There has been much interest by the natural resource sciences to lessen the challenges and develop new and better models for integrating both information and decision making across disciplines and ecosystems (Lockwood et al. 2010; Benson and Garmestani 2011). Also, there have been long-standing efforts to integrate learning from across the natural, social, and health sciences to understand and respond to the determinants of health in the context of wider processes of ecological and social change (see Chap. 5; also Rapport et al. 1998; McMichael et al. 1999). Efforts in the 1990s have fuelled a range of such integrative approaches associated, for example, with developments in ecosystem health, One Health, and ecosystem approaches to health (Charron 2012; Wilcox et al. 2012; Hallstrom et al. 2015). Calls for *sustainability sciences* to bridge the sociopolitical and ecological dynamics of sustainability (Kates et al. 2001) are further expressions of the ongoing scholarly and applied challenge of integration.

Informed by these precedents, calls for integrative approaches to cumulative impacts are inevitable and overdue. We note, however, that fostering integrative, cumulative thinking can be especially difficult as some impact types are counter cyclical. Following a decline in global economic activity, resource development may decrease, lessening the associated negative impacts in the environment. At the same time, communities might struggle to accommodate a decline in local economic investment, including reductions in tax revenue and loss of employment. Likewise, faltering economies can lessen human well-being and increase health-care costs (Houle and Light 2014).

Yet despite these challenges, the science of decision-support and related integrative precedents have advanced greatly over the past decade and can help reduce the complexity that is the integration of impacts (Schneider et al. 2003; Martin et al. 2009). We can now model the change in ecosystem services relative to economic activity that is measured using any number of indices. Although technical and information gaps will continue to confront multi-stakeholder decision-making processes that consider a range of conflicting values, this is not the primary challenge. From our perspective, integration is hindered most by conventional thinking and a failure to depart from traditional practice.

### 8.3.2 Cumulative Thinking: Shifting Our Gaze beyond Megaprojects and Across Multiple Scales

Cumulative thinking is about more than the consideration of the range of socioecological impacts (e.g. Partidário 2012). At the heart of this concept is the understanding that all changes and influences on our environment are cumulative resulting in additive, multiplicative, synergistic, interactive, and non-linear impacts. Thus, cumulative thinking requires the full consideration of all of the driving forces of change and their associated impacts, across multiple spatial and temporal scales. This includes the accumulation of both small and large impacts from the past as well as those that occur in the future. The potential array of interactive consequences from multiple drivers of change at small and large scales within the same landscape is a recurring theme that was demonstrated throughout the vignettes in this book (see Chap. 6).

While developing a new vision for identifying and addressing impacts, we have recognised the strong tendency toward a focus on large development projects, often masking the multiple small changes occurring within the same region (Chap. 7; Fig. 7.1). Understandably, the media and public interest are drawn to new multi-billion dollar mega-development projects that generate notable economic investment and potential risks to the environment—these are *cumulative impacts*. In Western Canada, such development projects appear ever more frequently, from large-scale hydroelectric projects to oil and natural gas pipelines to mines. This has led to an increasing bias in regulatory assessment and mitigation to consider just single large projects with a relatively high likelihood of significant impacts. With the goal of efficiency, legislative processes in Canada are moving more in this direction (Doelle 2012; Gibson 2012; Kirchhoff et al. 2013). The insidious nature of cumulative impacts, however, is the many small changes to landscapes that go unnoticed until large impacts become apparent. Many of these small changes receive no consideration within cumulative effects policy and legislation, falling below thresholds for consideration during EA (Johnson 2011). An effective framework must fully consider all sources of effects, from small to large development projects, or it has failed to address the true nature of cumulative impacts.

We are not stating that there is no role for prioritising the impacts from the range of development projects that may occur across a region. Prioritisation, however, should be based on the magnitude or extent of cumulative impacts across development sectors not the size of any one project. An important hallmark of cumulative thinking is the explicit consideration of the scalar and nested nature of cumulative impacts. This includes both the spatial and temporal dimensions of cumulative changes as well as variation in the magnitude of impacts. We must consider more than the present and more than just a single or small suite of large, easily identified projects. By definition, the spatial and temporal character of cumulative impacts demands us to shift our gaze across past, present, to future changes, and to explicitly consider interactions that zoom in and out from the micro through to regional and up to global scales. Making this difficult, the drivers of change can have an unanticipated spatial scope reaching beyond what might at first be considered the footprint of one or multiple effects. Water and air pollution are excellent examples of cumulative impacts that can result from multiple sources and extend over large areas and time periods in often unpredictable ways (see Box 3.3). A complete accounting of the scalar nature of cumulative impacts has both philosophical and technical implications that clearly challenge conventional assessment and regulatory approaches.

## 8.4 Principles for Implementing Cumulative Thinking

Cumulative thinking is the starting point for developing new approaches for assessing and addressing cumulative impacts. A new philosophical approach to cumulative impacts, however, offers little tangible direction to communities and regions struggling with accelerating industrial development. Thus, we provide six principles that ground cumulative thinking in the reality of an integrative regional cumulative impacts framework designed to progress a vision and actions for sustainability. These principles support five elements that serve as the building blocks for such region-specific frameworks (see Sect. 8.5; Fig. 8.1). Based on the voices presented in this book and previous research and practice from around the globe, we believe that all of the principles are essential. Depending on the types of challenges, rates of development, and existing progress toward sustainability, individual regions may prioritise the realisation of each principle.

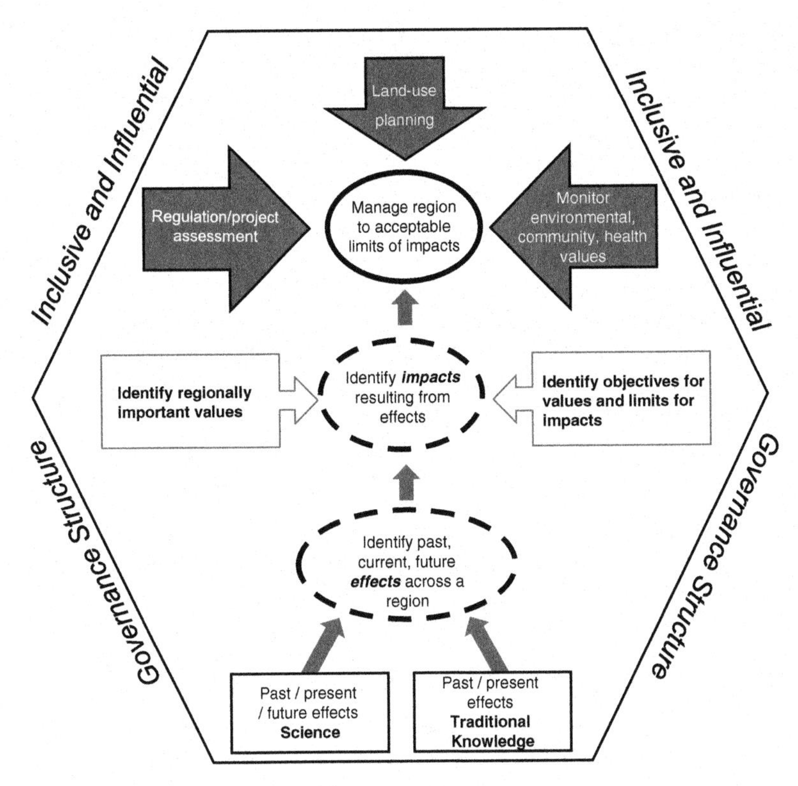


**Fig. 8.1** Conceptual integration of the necessary elements of an integrative regional cumulative impacts framework. Progression in the development and implementation of framework is represented from *bottom* to *top* by *ovals*. Key elements for development of the framework are presented in *rectangular boxes*, and *large green arrows* represent key elements for supporting the developed framework. Example of regionally important values could include water quality, employment, species at risk, and recreation. The *background hexagon* indicates the overriding importance of effective governance as well as the potential for the complementarity of cumulative impacts frameworks that may differ, but are adjacent across regions

### *Principle #1: Pursue Revolution not Evolution*

There is much evidence that current approaches for addressing cumulative impacts are ineffective (see Chap. 7). Such conclusions are at the core of our experience and the primary message of this book. As stated above, we can only solve this problem through a revolution in strategy not evolution of practice. Thus, we urge a new integrative way of thinking that is inclusive of ecosystem functioning, community well-being and balanced prosperity, and the health of those people that are affected by cumulative development across regions. A new way of thinking will require bold leadership and largely untested ways forward.

Some have experimented with regional approaches for cumulative impacts and these lessons will provide guidance (Harriman and Noble 2008). There is much room, however, for new ideas. Indeed, the failure of project-specific CEA and the lack of successful models for RSEA suggest we must build on the past, but investigate other approaches. The comfortable and risk-averse strategy for addressing cumulative impacts is to adjust, adapt, or evolve. For rapidly expanding or emerging resource economies, such change may be too slow.

We are not suggesting that new and radically different frameworks will be easy or free of failure. Integration is inherently difficult. The evidence is irrefutable, however: current methods for CEA and means of addressing cumulative impacts are not working (Duinker and Greig 2006).

***Principle #2: Context-Specificity***

Cumulative impacts are context-specific and particular to regional circumstances. Thus, each cumulative impacts framework must be crafted to meet the needs and particularities of the local ecology, affected communities, and the dictating circumstances of resource industries and economies. Although there is no simple formula for developing and implementing broad-scale and integrative frameworks for addressing cumulative impacts, we have learned much over the past 20 years that can be applied to the development of frameworks today (Harriman and Noble 2008). Many of those lessons, however, are related to what has not worked. This has implications for improved governance structures that focus on cumulative impacts in their entirety and strategies that can actively address the problem, not simply predict and report current and future impacts (Gunn and Noble 2009b). We are confident that efforts focused on bold approaches will result in new insights for addressing cumulative impacts across regional landscapes. Such experimentation will provide additional guidance some of which will serve as starting templates for the development of new frameworks as well as the adaptive improvement of contemporary efforts.

***Principle #3: Urgency of Action***

Through our experiences working in landscapes and communities with gradual or sudden increases in cumulative impacts we have come to appreciate the direct financial and political costs of failing to act in the present, instead delaying for the future. Indeed, we believe that communities confronting developing economies or relatively undeveloped landscapes should act to address cumulative impacts now. Where lands are publicly owned, such as Canada, governments have options to direct their future, including identifying the types of development that are suitable and the limits that might guide the placement or amount of future development.

Strategic planning, including the consideration of future cumulative impacts, is best accomplished when we have a full set of options that are not constrained by contractual rights to minerals, forests, or energy. When tenures and licences are sold or resource industries become entrenched through tax and employment benefits we limit the options for how we might manage a region (Timoney and Lee 2001). The importance of considering and weighing a full range of development futures today is especially important when setting limits for cumulative impacts. By their very nature, the impacts of many small cumulative effects are difficult to measure or predict. Although it may be possible to sequentially evaluate and address large mega-development projects, we are unlikely to be able to address each of the cumulative impacts that can occur at small scales and across multiple resource sectors. It is imperative that we consider the full range of known and anticipated cumulative effects today if we are to maintain a full set of relatively inexpensive decisions that will meet the vision of the future as expressed by potentially affected communities.

The urgency to develop a vision and framework for addressing regional cumulative impacts is exacerbated by the rapid development of global markets and the rush to establish long-term trade agreements focused on commodities. The first decades of the twenty-first century have, for example, witnessed a race between Canada, USA, Australia, and Russia to attract investment, develop the export facilities, and then lock-in long-term contracts to sell LNG to Asia. Some would argue that the reduction in process and mandatory review times for the *Canadian Environmental Assessment Act* (Doelle 2012; Government of Canada 2012) is designed to increase Canada's competitiveness in such negotiations. Regardless, the establishment of long-term trade agreements is likely to limit options to realise regional visions of the level of acceptable development and associated cumulative impacts.

#### *Principle #4: Inclusive Representation of Values*

Sustainability is a core tenet in the development of effective frameworks for addressing cumulative impacts (Cocklin 1993; Duinker and Greig 2006). Goals for sustainability are defined by values that are intended, ultimately, to represent the direct and indirect needs of people. To meet those needs, we must have healthy and functioning environments, economies, and communities. Thus, as a corollary to sustainability, we must develop inclusive processes for identifying the diverse values of communities and their supporting environments (Partidário 2012). The need for inclusive processes and dialogues about choices, and the implications of our choices, is further reinforced when we recognise that our definitions of sustainability will rarely be exactly the same. Inclusion can help ensure that the fullest range of values is identified. Those values are then translated into goals that are the starting point for strategic approaches for ensuring sustainable-levels of human activity within the context of both positive and negative cumulative impacts.

As highlighted throughout this book, the cumulative impacts of development can influence all forms of values from the spiritual and cultural connections to the land to ecosystem services to access to medical care (see Chap. 6). People, governments, and industry must be given an equal voice to express those values and to debate trade-offs in values (McShane et al. 2011). In some cases, those voices will be from beyond the region where cumulative impacts are directly expressed. State and national perspectives on economic development, biodiversity, and cultural well-being may also play a role in goal setting and ultimately decision making.

A more comprehensive accounting of values demands that we are inclusive of cultural perspectives and knowledge systems. Culture underlies values and will play an important role in defining the wants of communities and the trade-offs in positive and negative impacts. In BC and across Canada, the necessity of cross-cultural approaches is best highlighted by the unique economic aspirations and perspectives of Aboriginal communities (see Sect. 1.7). Indigenous communities in other parts of the world, for example the USA, New Zealand, and northern Russia, have similar and equally informative experiences (Dixon and Montz 1995; Kinnear et al. 2013).

Indigenous people bring a unique perspective on the workings of the natural world and the role of people in that world. As part of that cultural dynamic, knowledge systems other than science might help understand or interpret development impacts. As one example, TEK can reveal or complement science in documenting long-term patterns in natural disturbance or the distribution of plants and animals (Wenzel 1999; Berkes et al. 2000; Santomauro et al. 2012). Similarly, relatively long time series of information can be offered by Local Ecological Knowledge (Azzurro et al. 2011). Thus, we envision inclusivity as more than a seat at a table, but the recognition of deep-rooted cultural perspectives that bring necessary breadth to the range of values we might consider and new insights to our understanding and the management of cumulative impacts.

***Principle #5: Adaptive and Iterative Frameworks***

Addressing cumulative impacts is a wicked problem (see Chap. 7). In recognising this, we are especially challenged to acknowledge the risk of unintended consequences: whereby efforts to improve the current system have the potential to create new problems (Brown et al. 2010; Hallstrom et al. 2015). Indeed, our proposed approach to cumulative thinking, although essential for addressing the problem, increases the level of wickedness. As a wicked problem there is no definitive a priori formula for solving cumulative impacts or even a general measure of success. This demands an explicit recognition to ongoing learning, rather than assuming that it is possible to propose a final, correct strategy. Cumulative impacts within each regional setting are unique and dynamic depending on anticipated and unanticipated changes to local and national economies as well as the ecosystems that support those economies (see Chap. 6). In addition, moving from a focus on environmental change to integrating community and health values makes for a much more complex set of impacts and interactions. Likewise, consideration of impacts over broad scales of time, space, and magnitude make quantification difficult and greatly increases the uncertainty in predicting future outcomes. Any framework designed to address cumulative impacts will need to be adaptive to a changing region, including the changing environment and the changing needs of the people that inhabit that region.

Given the high levels of uncertainty that underlie landscape-level ecosystems and economies, an effective framework will need to explicitly recognise gaps in knowledge, and incorporate mechanisms for filling those gaps and applying that new understanding. Adaptive management is one approach for focused and applied knowledge generation that can be implemented at the relatively large spatial scales that characterise cumulative impacts frameworks (Walters 1986; Allan and Stankey 2009). The limitations of adaptive management, however, need to be recognised and addressed (McLain and Lee 1996; Simberloff 1998). Furthermore, the emphasis has to be knowledge application not just knowledge generation (Bammer 2005; Westgate et al. 2013). We must note that new understanding can demand sweeping departures from current practices, processes, or even management paradigms. There must be a willingness to accept and implement radical change, and to commit to ongoing processes of (re)learning and change.

***Principle #6: Transparency of Process***

Inclusiveness is only fully realised if governance structures allow for the involvement of representative voices. Full participation of people in a region is best facilitated by clearly and openly stating the goals and objectives of a cumulative impacts framework as well as the process for identifying, achieving, measuring, and potentially modifying those objectives. As with the other principles, transparency will need to be crafted on a region by region basis. There is some guidance, however, that can be taken from existing community-based management systems with cross-cultural co-management offering one such example (Tipa and Welch 2006; Armitage et al. 2011).

To be successful, communication and engagement in the discussion of the types and levels of resource development needs to occur across the range of demographic, socioeconomic, and cultural sectors of the community (Lawrence and Deagen 2001; Parkes 2015). This means that *outreach* must include more than the direct stakeholder organisations or their representatives within some structured framework. Ultimately, transparency of process demands engagement of members of the affected communities in vision, goal, and objectives setting, implementation of strategies, and perhaps monitoring and knowledge production (CCME 2009). Knowledge sharing is a requirement for transparency and an effective cumulative impacts framework (Noble et al. 2013). Citizen science is one example of a process that allows direct involvement in monitoring impacts, perhaps reducing costs and increasing ownership and trust in locally relevant knowledge (Cohn 2008; Ottinger 2009; Riesch and Potter 2014). Given the unique nature of most regions that are struggling with cumulative impacts, inclusiveness facilitated by transparent processes is essential for finding effective solutions. Again, referring to the wickedness of this problem, the individual people, families, and organisations struggling with cumulative impacts are best suited to identify the unique and relevant elements of the issue and work within a structured framework to identify solutions (see Vignette 6.7).

## 8.5 Elements of a Locally Adaptable Integrative Regional Cumulative Impacts Framework

We do not recommend a single one-type fits all framework for addressing cumulative impacts. Across Canada, REA has taken many forms with each being designed or perhaps simply evolving to meet regional circumstances and challenges (Harriman and Noble 2008). Based on the work presented in this book, and drawing from experiences in other jurisdictions, we believe that there are a number of general elements that are essential for the success of an integrative regional cumulative impacts framework. Indeed, systematic studies of the successes and failures of REA have focused on general recommendations not the details of building and applying such frameworks (e.g. Gunn and Noble 2009a). We have identified five elements that we see as necessary components of an effective cumulative impacts framework; although, we provide only general descriptions to ensure that they are adapted for each unique circumstance. Each element can be linked to the six principles that support cumulative thinking.

As a starting point for developing a cumulative impacts framework, we provide a structure for how these various elements might be integrated (see Fig. 8.1). This depiction should, however, be seen as an expression of a hypothesis, rather than a conclusive proposal. We challenge communities, regions, and governments to adapt the principles of cumulative thinking and connect the various elements as appropriate for their circumstances. Recognising the principle of adaptation, we caution that even within a particular region such a framework may need to change and improve over time. This may include modifying the form or structure of any one element as well as the interconnectedness of the elements that constitute the framework.

***Element #1: Access to Timely and Appropriate Knowledge***

At the heart of rational and transparent decision making is knowledge (Conroy and Peterson 2013) and the dynamics of knowledge generation, application, translation, and exchange (Allen et al. 2001; Roux et al. 2006). This is especially the case when considering the complexity of cumulative impacts including the multiple spatial and temporal scales over which they occur. Understanding change is dependent on knowledge that represents the biophysical attributes of the environment as well as the social and cultural dimensions of dependent communities. This will include a comprehensive accounting of existing effects and impacts and the drivers of change, including their interactions. Such baseline knowledge is essential for recognising and measuring an increase in the magnitude, extent, trajectory, or the emergence of new impacts. There must, however, be a willingness to share knowledge that will allow diverse and non-affiliated groups to collectively address the challenges of cumulative impacts (Noble et al. 2013). Essential gaps in the knowledge of the processes underlying change or the inventory data necessary for measuring and assessing impacts and effects must be addressed collectively.

Knowledge has many elements, sources, and, in some cases, a cultural context that must be recognised if we are to ensure inclusivity. Furthermore, knowledge production can be a powerful tool for building collaborative and inclusive decision making when it reaches beyond the domain of the expert (Parkes et al. 2005; Pohl 2008). Community-based monitoring, for example, can help identify impacts and trends in both the biophysical and human components of the environment while improving relationships and broadening communication between industry and affected communities (Noble and Birk 2011). Developing meaningful, relevant, and sustainable monitoring systems can be challenging, requiring considerable resources to ensure the application of effective design principles and the outreach and application of findings (Whitelaw et al. 2003; Luzar et al. 2011).

Some frameworks and tools for addressing biophysical impacts have emphasised prediction (Schneider et al. 2003; see Chap. 3). When coupled with scenario-type decision frameworks this can be a powerful tool for exploring uncertain futures (Peterson et al. 2003; Weber et al. 2012). Also, a range of scenarios depicting future development, and the positive and negative impacts, can act as effective prompts for helping communities to explore and develop a vision for their region. Some have reported, however, that an overemphasis on the technical aspects of impacts, including modelling, can be a distraction from governance and decision making (Gunn and Noble 2009b). Indeed, models, no matter how sophisticated and accurate, cannot make decisions or set priorities they can only provide quantitative estimates of change.

### *Element #2: Identifying and Measuring Agreed-Upon Values*

Identifying and measuring values is an essential element of an effective cumulative impacts framework. It can be difficult, however, to meaningfully identify and represent a community's or region's values. The concept of criteria and indicators offers some guidance for structuring processes that are effective at identifying and then monitoring the success of a planning framework or management system for maintaining values (Axelsson et al. 2013; Steenberg et al. 2013).

When considering a large spectrum of values, from citizen health to ecosystem services, there is likely some benefit to refining and maintaining a variation of the existing VEC structure, as is currently implemented for project-specific EA (Gunn and Noble 2009a). Although reductionist, prioritisation is an unescapable reality of natural resource and conservation management and planning (Hirsch et al. 2011). Taking into account that the broader integration of environment, community, and health is central to the intent of cumulative thinking, we propose a shift to consider valued socioecological components in place of the existing focus on VECs (see Chap. 2). Following on the work of Gunn and Noble (2009a), we also recommend choosing regionally representative, as compared to species- or area-focused, valued socioecological components. For example, an increase in the density of roads across a region may be an indirect measure of a change in the persistence and distribution of wildlife, a reduction in opportunities for nature-based tourism, and a general decrease in the aesthetic values associated with unroaded landscapes. Drawing from best practices for identifying effective criteria and indicators, one might also choose individual values that resonate across the environment, communities, and health. Water quality, availability, and access, for example, are of universal importance to human and ecological communities.

### *Element #3: Commitment to Monitoring and Assessment*

Following the identification of values and associated indicators, processes must be designed and implemented to determine if the objectives for those values—increase, decrease, or maintain status quo—are being met. Thus, measurement and likely some form of long-term monitoring is an essential element of any adaptive framework designed to assess and manage cumulative impacts (Dubé et al. 2006; Burton et al. 2014). Without monitoring it is difficult to understand or predict the type or rate of impacts, or to learn about the long-term impacts of prior changes. Furthermore, monitoring is an essential component of formal adaptive management experimentation (Walters 1986). Observed change in the quality, distribution, or availability of some value, for example health care or aesthetics of the environment, can potentially change management direction, trigger a management response or inform future decisions.

Although it is relatively easy to identify a long list of measurable attributes of values, such measures can be difficult to implement especially if they are costly or technically challenging. Furthermore, imprecise measures are ineffective for assessing change or trends in the quantity, quality, or general state of some value over time. Where community involvement is a key attribute of a successful cumulative impacts framework, monitoring protocols should be accessible and inclusive to a broad-spectrum of participants. This will help ensure the sustainability of monitoring programmes, as community members are engaged and invested in success (Conrad and Daoust 2008; Fernandez-Gimenez et al. 2008). Furthermore, a broad understanding and ownership of such information will increase the likelihood that the results of monitoring programmes are relevant to communities and that they are applied to inclusive cumulative impacts frameworks (O'Faircheallaigh 2007). Scientific measurement is normally the domain of experts. Thus, many monitoring programmes are designed and implemented by technocrats using methods that are unapproachable for the untrained. This may work for some questions or scales of measurement, but will have limited success within an integrative regional cumulative impacts framework that is designed to be inclusive of many voices and perspectives.

### *Element #4: Strategic Planning and Decision-Making Frameworks*

We add our voice to those scholars and practitioners who have stated that solutions for cumulative impacts at regional scales must incorporate elements of strategic land-use planning, and that planning and decision making will need to be contextualised within choices about governance processes (Bardecki 1990; Booth and Skelton 2011; Johnson 2011). This is especially the case when considering multi-sector REA and management (Chetkiewicz and Lintner 2014). Adopting a land-use planning approach would: ensure the consideration of a broad range of perspectives and the involvement of all stakeholders and regulatory agencies; identify a range of effects and impacts early in the decision-making process; assess baseline conditions and identify knowledge gaps; and develop a holistic vision for cumulative impacts including the identification of key values and supporting goals for those values. Importantly, strategic planning would structure decision making so that it is collaborative and forward looking, potentially representing large regional areas and time periods, but including mechanisms for tracking progress toward meeting goals (Jackson 2002; Frame et al. 2004). Here, the role of monitoring and adaptive management would be essential. Failure to consider large spatial and temporal scales and to adapt to new circumstances is one of the inherent limitations of considering cumulative impacts within reactive and place-specific EA.

As part of the strategic vision of land use for a region, members of a planning team would set goals and objectives for levels of future development, including benefits and services to local communities, as well as the conservation of ecological values and ecosystem services. Strategies would be developed to meet that set of objectives for each value. Within an integrative regional cumulative impacts framework, we envision targets or regulatory limits to ensure that the effects and impacts do not exceed predefined levels that might compromise the quality or availability of a value.

Regulatory limits on development often are referred to as thresholds; although, there is considerable confusion and debate about what might constitute a threshold and even less certainty about how to measure complex thresholds, especially those that relate to ecosystem function (Johnson 2013). Furthermore, thresholds based exclusively on biophysical parameters are often coopted or compromised by socio-economic considerations. Thus, we recommend the application of regulatory limits that explicitly balance ecosystem function with economic activity. Such trade-offs represent risk–reward relationships that are probabilistic in nature. Increasing cumulative effects will result in a higher probability of risk for some component of the environment or other human value. Given the uncertainty in such risk-reward relationships, the exact magnitude or probability of occurrence of risk is often unknown. This is especially the case for acute or catastrophic impacts that cannot be monitored for trends in change; for example an oil spill or the failure of a tailings pond. Likewise the failure to accept some risk could forego economic opportunity for communities. As an example, federal guidance for the conservation of woodland caribou, a protected species in Canada, recommends a limit of 35 % cumulative disturbance across a herd's range; that level of impact is thought to result in a 60 % probability of the population persisting (Environment Canada 2012). An increase in disturbance would decrease the probability of persistence, but likely result in greater economic development. Monitoring and research can help inform such risk-reward relationships potentially refining or resetting regulatory limits for certain activities that may result in cumulative impacts for important values.

The setting of binding regulatory limits is a key and differentiating element of our proposed framework relative to similar approaches that have been implemented in Canada. Other frameworks have decision-making structures that at best allow *refusable advice* to governments that may then act on development proposals according to proposed or accepted limits (Johnson 2011). Meaningful limits, even if accommodating trade-offs in values, should represent stopping points for development and associated cumulative impacts within a region. Firm limits would also provide guidance to higher levels of government and project proponents. Some have criticised limits as they are viewed as a constraint on economic development (Salmo Consulting 2006). We see limits as a way of representing and realising a vision that potentially reduces uncertainty for project proponents that are forced to navigate unpredictable and expensive EA processes.

### *Element #5: Transparent and Inclusive Governance Regimes*

Governance is different from government. In keeping with earlier chapters, we describe governance as "a process whereby societies or organisations make their important decisions, determine whom they involve in the process and how they render account" (Graham et al. 2003, p. 1). The integrative aspect of governance is multifaceted (Parkes et al. 2010). One type of integration is horizontal through processes that connect across similar groups. Determining who will be involved in horizontal integration raises questions about how inter-sectoral and interdisciplinary our processes are, in ways that are often challenged by entrenched organisational structures and knowledge cultures. Integrative governance and decision making also involves vertical integration that links among different types of knowledge and participation to facilitate understanding and action (Parkes et al. 2005). When assessing cumulative impacts, vertical integration would recognise the value and contributions of scientific, local, and TEK systems. Governance provides the decision-making framework within which to recognise values, identify goals and objectives, and implement strategies to achieve those higher-level ambitions.

Without effective governance, the principles of inclusivity, adaptation, and transparency are lost or not fully realised. Achieving effective governance requires leadership and commitment by participating people and organisations. It also requires that those participating people and organisations are able to work together effectively, develop functional networks, and a level of trust—in other words they need to develop and deploy social cohesion that bridges and bonds social capital (Noble et al. 2013). Indeed, by extending beyond traditional government processes in order to address *who* is involved and *how* we make decisions our choices about governance determine options for many of the other principles underlying an integrative cumulative impacts framework.

When looking for useful, tangible, precedents for integrative governance processes, some guidance may be taken from structures that have been applied to other cumulative effects frameworks that have been implemented across Canada. As an example, the Cumulative Environmental Management Association, operating across northern Alberta, consists of over 50 members that are placed within one of four caucuses: Aboriginal, Government, Nongovernment Organisations, and Industry (see: http://cemaonline.ca/). The Association is governed by four directors from each caucus. This group is charged with formulating management frameworks, best practices, and integrated dialogue with the purpose of reducing cumulative impacts, largely as a result of oil sands development. Their governance structure allows for recommendations to the federal and provincial government with no powers for actual decision making.

We believe that some form of autonomous decision-making body or authority will be a key element of effective governance for addressing cumulative impacts across regions. This would set new approaches apart from existing examples. To start, few jurisdictions have regional government structures of a size or alignment (i.e. watershed) that would be effective to these purposes, and in those that do the decision-making powers tend to be weak. Most current frameworks in Canada, including past experiences with strategic planning, are tasked only with an advisory role. Advice has value, but can be disregarded by governments depending on broader provincial or federal priorities (Halseth and Booth 2003). Also, the relevance and importance of a framework, and ultimately participation in that framework, would be undermined if government disregarded advice and failed to act on key issues such as regionally identified limits on development or objections to large projects (Mascarenhas and Scarce 2004).

## 8.6 Realising an Integrative Regional Cumulative Impacts Framework

The six principles and five elements of what we consider cumulative thinking are not meant as a solution or replacement for all natural resource, community, or health decision making. There are other processes for review and practices designed to minimise the impacts of resource development on ecosystems and communities. The challenges of economic diversification and rural health require a consideration of the cumulative impacts of resource development, but there are many other elements that have little relevance to human-caused changes to the environment. Even when considering environmental protection and management, there will need to be approvals, licences, and tenures that are informed by levels and limits of cumulative impacts, but that are largely site-specific decisions.

When considering resource development, some have suggested a continued role for project-specific review, mitigation, and approval. Such processes would be tiered or placed in the context of a broader cumulative impacts vision and targets for larger regional areas. Thus, there needs to be both strategic and operational decision making (Gunn and Noble 2009a). A cumulative impacts framework may be too coarse of a tool for considering the review and approval of large mega-projects. This type of industrial development can have significant impacts that are ultimately cumulative, but must be considered relative to the benefits and risks of the individual project. Furthermore, the incremental impacts of such development projects may be diluted when considered across a larger regional study area, thus site specificity is still important (MacDonald 2000). Also, such projects may cross administrative boundaries or have wide-ranging impacts (e.g. oil pipeline), requiring an even broader-scale of public involvement and review. For many legislated review processes (e.g. *BC Environmental Assessment Act*; Government of BC 2002), development thresholds trigger an assessment. Thus, at the opposite end of the development-magnitude spectrum, many projects do not fall within the scope of conventional EA. Furthermore, the less conventional impacts, including those to communities and indirect determinants of health, are often not considered during the review process. Environmental assessment is valid for some development projects for a limited set of values, but an integrative cumulative impacts framework is essential if we are to fill the holes created by conventional practice.

We recommend working at the scale of the region when applying the five elements of the proposed cumulative impacts framework (Fig. 8.1). This is consistent with current examples and conceptual thinking that has forwarded the idea of REA as a more effective approach for managing cumulative impacts (Bonnell and Storey 2000; Dubé 2003; Gibson et al. 2010; Chetkiewicz and Lintner 2014). A regional study area should be defined in a way that is meaningfully informed by context, taking into account current political boundaries, biophysical attributes such as watersheds, the distribution of a particular resource sector and associated effects, or the boundaries of communities facing like challenges. In contrast to project-specific assessments that define the study area according to the spatial extent of the stressor, a cumulative thinking approach would consider the broader region that represents both the processes and the players that influence resource use or are affected by the development of those resources (MacDonald 2000).

We must emphasise the necessity of crafting a cumulative impacts framework that is tailored for the unique development challenges, economic opportunities, community aspirations, and ecology of an appropriately defined regional area. As recommended by Harriman and Noble (2008, p.27) the choice of a framework must be predicated on the necessity that it is "fit-for-purpose", not imposed in a way that cannot respond to the unique social or ecological characteristics of the region. Thus, the elements and principles described here are not overly complex or prescriptive. Any one highly structured framework may be too complex or perhaps inconsistent with a particular region's visions or cumulative impacts challenges. Such guidance would have a high potential of producing stagnation and status quo thinking, not steps forward.

We recognise that there is little formal government or institutional structure to support a fully implemented framework. In BC, as in most of Canada, resource management policy is centralised; although there was a time in the recent past where strategic planning was delegated to regional and inclusive planning tables (Cashore et al. 2001; Jackson and Curry 2002). Unfortunately, there are no obvious signs or examples that governments are willing to cede decision making to regional bodies. This approach to governance is increasingly frustrating citizens and communities that feel there are too few avenues for dialogue and little opportunity to influence the current rates and types of natural resource development. The voices and experiences represented in this book suggest that the gulf between development and communities, including the associated challenges and impacts, will only increase. Other emerging resource economies around the world likely face similar challenges (Partidário 2012).

Application of the principles and elements described here is a starting point for addressing such disconnect between people and their environments. Much of the guidance can be applied incrementally by building discussion, then dialogue with decision-makers, and finally decision-making frameworks with well-stated goals and objectives and effective governance structures. We believe that such approaches have the potential to better serve a range of interests, meeting the needs and wants of communities, allowing for consistent processes of review and approval for industry, and more democratic and predictable land-use decision making for government.

## References


Allan, C., and G.H. Stankey (eds.). 2009. *Adaptive environmental management—A practitioner's guide*. Collingwood: CSIRO Publishing.

Allen, W., O. Bosch, M. Kilvington, D. Harley, and I. Brown. 2001. Monitoring and adaptive management: Resolving social and organisational issues to improve information sharing in natural resource management. *Natural Resources Forum* 25: 225–233.

Armitage, D., F. Berkes, A. Dale, E. Kocho-Schellenberg, and E. Patton. 2011. Co-management and the co-production of knowledge: Learning to adapt in Canada's Arctic. *Global Environmental Change* 21: 995–1004.

Axelsson, R., P. Angelstam, E. Degerman, S. Teitlbaum, K. Andersson, M. Elbakidze, and M.K. Drotz. 2013. Social and cultural sustainability: Criteria, indicators, verifier variables for measurement and maps for vizualisation to support planning. *Ambio* 42: 215–228.

Azzurro, E., P. Moschella, and F. Maynou. 2011. Tracking signals of change in Mediterranean fish diversity based on local ecological knowledge. *PLoS ONE* 6(9): e24885.

Bammer, G. 2005. Integration and implementation sciences: Building a new specialization. *Ecology and Society* 10(2): 6.

Bardecki, M.J. 1990. Coping with cumulative impacts: An assessment of legislative and administrative mechanisms. *Impact Assessment Bulletin* 8: 319–344.

Barnes, T.J., and M. Gertler (eds.). 1999. *The new industrial geography: Regions, regulation and institutions*. New York: Routledge.

Benson, M.H., and A.S. Garmestani. 2011. Can we manage for resilience? The integration of resilience thinking into natural resources management in the United States. *Environmental Management* 48: 392–399.

Berkes, F., J. Colding, and F. Carl. 2000. Rediscovery of traditional ecological knowledge as adaptive management. *Ecological Applications* 10: 1251–1262.

Bonnell, S., and K. Storey. 2000. Addressing cumulative effects through strategic environmental assessment: A case study of small hydro development in Newfoundland, Canada. *Journal of Environmental Assessment Policy and Management* 2: 477–499.

Booth, A.L., and N.W. Skelton. 2011. Improving First Nations' participation in environmental assessment processes: Recommendations from the field. *Impact Assessment and Project Appraisal* 29: 49–58.

Brown, V.A., J.A. Harris, and J.Y. Russell. 2010. *Tackling wicked problems: Through the transdisciplinary imagination*. London: Earthscan.

Brundtland, G.H., and World Commission on Environment and Development. 1987. *Our common future: Report of the World Commission on Environment and Development*. Oxford: Oxford University Press.

Burton, A.C., D. Huggard, E. Bayne, J. Schieck, P. Sólymos, T. Muhly, D. Farr, and S. Boutin. 2014. A framework for adaptive monitoring of the cumulative effects of human footprint on biodiversity. *Environmental Monitoring and Assessment* 186: 3605–3617.

Cashore, B., G. Hoberg, M. Howlett, J. Rayner, and J. Wilson. 2001. *In search of sustainability: British Columbia forest policy in the 1990s*. Vancouver: University of British Columbia Press.

CCME (Canadian Council of Ministers of the Environment). 2009. *Regional strategic environmental assessment in Canada: Principles and guidance*. Winnipeg: Canadian Council of Ministers of the Environment.

Charron, D.F. 2012. *Ecohealth research in practice: Innovative applications of an ecosystem approach to health*. New York: Springer.

Chetkiewicz, C., and A.M. Lintner. 2014. *Getting it right in Ontario's north: The need for regional strategic environmental assessment in the Ring of Fire [Wawangajing]*. Toronto: Wildlife Conservation Society and Ecojustice.

Cocklin, C. 1993. What does cumulative effects analysis have to do with sustainable development? *Canadian Journal of Regional Science* 16: 453–479.

Cohn, J.P. 2008. Citizen science: Can volunteers do real research? *BioScience* 58: 192–197.

Conrad, C., and T. Daoust. 2008. Community-based monitoring frameworks: Increasing the effectiveness of environmental stewardship. *Environmental Management* 41: 358–366.

Conroy, M.J., and J.T. Peterson. 2013. *Decision making in natural resource management: A structured, adaptive approach*. New York: Wiley-Blackwell.

Dixon, J., and B.E. Montz. 1995. From concept to practice: Implementing cumulative impact assessment in New Zealand. *Environmental Management* 19: 445–456.

Doelle, M. 2012. CEAA 2012: The end of federal EA as we know it? *Journal of Environmental Law and Practice* 24: 1–17.

Dubé, M., B. Johnson, G. Dunn, J. Culp, K. Cash, K. Munkittrick, I. Wong, K. Hedley, W. Booty, D. Lam, O. Resler, and A. Storey. 2006. Development of a new approach to cumulative effects assessment: A northern river ecosystem example. *Environmental Monitoring and Assessment* 113: 87–115.

Dubé, M.G. 2003. Cumulative effect assessment in Canada: A regional framework for aquatic ecosystems. *Environmental Impact Assessment Review* 23: 723–745.

Duinker, P.N., and L.A. Greig. 2006. The impotence of cumulative effects assessment in Canada: Ailments and ideas for redeployment. *Environmental Management* 37: 153–161.

Duinker, P.N., E.L. Burbidge, S.R. Boardley, and L.A. Greig. 2012. Scientific dimensions of cumulative effects assessment: Toward improvements in guidance for practice. *Environmental Review* 21: 40–52.

Environment Canada. 2012. *Recovery strategy for the woodland caribou (Rangifer tarandus caribou), boreal population, in Canada*. Species at risk act recovery strategy series. Ottawa: Environment Canada.

Esteves, A.M., D. Franks, and F. Vanclay. 2012. Social impact assessment: The state of the art. *Impact Assessment and Project Appraisal* 30: 34–42.

Fernandez-Gimenez, M., H. Ballard, and V. Sturtevant. 2008. Adaptive management and social learning in collaborative and community-based monitoring: A study of five community-based forestry organizations in the western USA. *Ecology and Society* 13(2): 4.

Frame, T.M., T. Gunton, and J.C. Day. 2004. The role of collaboration in environmental management: An evaluation of land and resource planning in British Columbia. *Journal of Environmental Planning and Management* 47: 59–82.

Gibson, R.B. 2012. In full retreat: The Canadian government's new environmental assessment law undoes decades of progress. *Impact Assessment and Project Appraisal* 30: 179–188.

Gibson, R.B., H. Benevides, M. Doelle, and D. Kirchhoff. 2010. Strengthening strategic environmental assessment in Canada: An evaluation of three basic options. *Journal of Environmental Law and Practice* 20: 175–211.

Government of BC. 2002. *Environmental Assessment Act, 2002* (S.B.C. 2002, C. 43). http://www.bclaws.ca/civix/document/id/complete/statreg/02043_01. Accessed 30 Mar 2015.

Government of Canada. 2012. *Canadian Environmental Assessment Act, 2012* (S.C. 2012, C. 19, S. 52). http://laws-lois.justice.gc.ca/eng/acts/c-15.21/page-1.html. Accessed 12 Dec 2014.

Graham, J., B. Amos, and T. Plumptre. 2003. *Principles for good governance in the 21st century. Policy brief no. 15—August 2003*. Ottawa: Institute on Governance. http://iog.ca/wp-content/uploads/2012/12/2003_August_policybrief15.pdf. Accessed 21 Nov 2014.

Gunn, J.H., and B.F. Noble. 2009a. A conceptual basis and methodological framework for regional strategic environmental assessment (R-SEA). *Impact Assessment and Project Appraisal* 27: 258–270.

———. 2009b. Integrating cumulative effects in regional strategic environmental assessment frameworks: Lessons from practice. *Journal of Environmental Assessment Policy and Management* 11: 267–290.

Hallstrom, L.K., M.W. Parkes, and N. Guelstorf. 2015. Convergence and diversity: Integrating encounters with health, ecological and social concerns. In *Ecosystems, society and health: Pathways through diversity, convergence and integration*, ed. L.K. Hallstrom, N. Guelstorf, and M.W. Parkes. Montreal: McGill-Queens University Press.

Halseth, G., and A. Booth. 2003. 'What worked well—what needs improvement'—Lessons in public consultation from British Columbia's resource planning processes. *Local Environment* 8: 437–455.

Harriman, J.A., and B.F. Noble. 2008. Characterizing project and strategic approaches to regional cumulative effects assessment in Canada. *Journal of Environmental Assessment Policy and Management* 10: 25–50.

Harris-Roxas, B., V. Francesca, A. Bond, B. Cave, M. Divall, P. Furu, P. Harris, M. Soeberg, A. Wernham, and M. Winkler. 2012. Health impact assessment: The state of the art. *Impact Assessment and Project Appraisal* 30: 43–52.

Hirsch, P.D., W.M. Adams, J.P. Brosius, A. Zia, N. Bariola, and J.L. Dammert. 2011. Acknowledging conservation trade-offs and embracing complexity. *Conservation Biology* 25: 259–264.

Houle, J.N., and M.T. Light. 2014. The home foreclosure crisis and rising suicide rates, 2005 to 2010. *American Journal of Public Health* 104: 1073–1079.

Jackson, T. 2002. Consensus processes in land use planning in British Columbia: The nature of success. *Progress in Planning* 57: 1–90.

Jackson, T., and J. Curry. 2002. Regional development and land use planning in rural British Columbia: Peace in the woods? *Regional Studies* 36: 439–443.

Johnson, C.J. 2011. Regulating and planning for cumulative effects: The Canadian experience. In *Cumulative effects in wildlife management: Impact mitigation*, ed. P.R. Krausman and L.K. Harris, 29–46. Boca Raton: CRC Press.

———. 2013. Identifying ecological thresholds for regulating human activity: Effective conservation or wishful thinking? *Biological Conservation* 168: 57–65.

Kates, R.W., W.C. Clark, R. Corell, J.M. Hall, C.C. Jaeger, I. Lowe, J.J. McCarthy, H.J. Schellnhuber, B. Bolin, N.M. Dickson, S. Faucheux, G.C. Gallopin, A. Grubler, B. Huntley, J. Jager, N.S. Jodha, R.E. Kasperson, A. Mabogunje, P. Matson, H. Mooney, B. Moore III, T. O'Riordan, and U. Svedin. 2001. Sustainability science. *Science* 292: 641–642.

Kennett, S.A. 2002. Lessons from Cheviot: Redefining government's role in cumulative effects assessment. In *Cumulative environmental effects management: Tools and approaches*, ed. A.J. Kennedy, 17–29. Calgary: Alberta Society of Professional Biologists.

Kinnear, S., Z. Kabir, J. Mann, and L. Bricknell. 2013. The need to measure and manage the cumulative impacts of resource development on public health: An Australian perspective. In *Current topics in public health*, ed. A. Rodriguez-Morales, 125–144. Rijeka: InTech Publishers.

Kirchhoff, D., H.L. Gardner, and L.J.S. Tsuji. 2013. The Canadian Environmental Assessment Act, 2012 and associated policy: Implications for Aboriginal people. *The International Indigenous Policy Journal* 4(3): 14–16.

Lawrence, R.L., and D.A. Deagen. 2001. Choosing public participation methods for natural resources: A context-specific guide. *Society and Natural Resources* 14: 857–872.

Lockwood, M., J. Davidson, A. Curtis, E. Stratford, and R. Griffith. 2010. Governance principles for natural resource management. *Society and Natural Resources* 23: 986–1001.

Luzar, J.B., K.M. Silvius, H. Overman, S.T. Giery, J.M. Read, and J.M.V. Fargoso. 2011. Large-scale environmental monitoring by Indigenous peoples. *Bioscience* 61: 770–782.

MacDonald, L. 2000. Evaluating and managing cumulative effects: Process and constraints. *Environmental Management* 26: 299–315.

Markey, S., G. Halseth, and D. Manson. 2012. *Investing in place: Economic renewal in northern British Columbia*. Vancouver: UBC Press.

Martin, J., M.C. Runge, J.D. Nichols, B.C. Lubow, and W.L. Kendall. 2009. Structured decision making as a conceptual framework to identify thresholds for conservation and management. *Ecological Applications* 19: 1079–1090.

Mascarenhas, M., and R. Scarce. 2004. "The intention was good": Legitimacy, consensus-based decision making, and the case of forest planning in British Columbia, Canada. *Society and Natural Resources* 17: 17–38.

McLain, R.J., and R.G. Lee. 1996. Adaptive management: Promises and pitfalls. *Environmental Management* 20: 437–448.

McMichael, A. 2006. Population health as the "bottom line" of sustainability: A contemporary challenge for public health researchers. *European Journal of Public Health* 16: 579–581.

McMichael, A.J., B. Bolin, R. Costanza, G.C. Daily, C. Folke, K. Lindahl-Kiessling, E. Lindgren, and B. Niklasson. 1999. Globalization and the sustainability of human health: An ecological perspective. *Bioscience* 49: 205–210.

McShane, T.O., P. Hirsch, T. Chi Trung, A. Songorwa, A. Kinzig, B. Monteferri, D. Mutekanga, H. Van Thang, J.L. Dammert, M. Pulgar-Vidal, M. Welch-Devine, J.P. Brosius, P. Coppolillo, and S. O'Connor. 2011. Hard choices: Making tradeoffs between biodiversity conservation and human well-being. *Biological Conservation* 144: 966–972.

Mebratu, D. 1998. Sustainability and sustainable development: Historical and conceptual review. *Environmental Impact Assessment Review* 18: 493–520.

Noble, B., and J. Birk. 2011. Comfort monitoring? Environmental assessment follow-up under community-industry negotiated environmental agreements. *Environmental Impact Assessment Review* 31: 17–24.

Noble, B.F. 2009. Promise and dismay: The state of strategic environmental assessment systems and practices in Canada. *Environmental Impact Assessment Review* 29: 66–75.

Noble, B.F., J.S. Skwaruk, and R.J. Patrick. 2013. Toward cumulative effects assessment and management in the Athabasca watershed, Alberta, Canada. *The Canadian Geographer* 58: 315–328.

O'Faircheallaigh, C. 2007. Environmental agreements, EIA follow-up and Aboriginal participation in environmental management: The Canadian experience. *Environmental Impact Assessment Review* 27: 319–342.

Ottinger, G. 2009. Buckets of resistance: Standards and the effectiveness of citizen science. *Science, Technology and Human Values* 35: 244–270.

Parkes, M.W. 2015. 'Just Add Water': Dissolving barriers to collaboration and learning for health, ecosystems and equity. In *Ecosystems, society and health: Pathways through diversity, convergence and integration*, ed. L.K. Hallstrom, N. Guelstorf, and M.W. Parkes. Montreal: McGill-Queens University Press.

Parkes, M.W., K.E. Morrison, M.J. Bunch, L.K. Hallstrom, R.C. Neudoerffer, H.D. Venema, and D. Waltner-Toews. 2010. Towards integrated governance for water, health and social–ecological systems: The watershed governance prism. *Global Environmental Change* 20: 693–704.

Parkes, M.W., L. Bienen, J. Breilh, L.-N. Hsu, M. McDonald, J.A. Patz, J.P. Rosenthal, M. Sahani, A. Sleigh, D. Waltner-Toews, and A. Yassi. 2005. All hands on deck: Transdisciplinary approaches to emerging infectious disease. *EcoHealth* 2: 258–272.

Partidário, M.R. 2012. *Strategic environmental assessment better practice guide: Methodological guidance for strategic thinking in SEA*. Lisbon: Portuguese Environment Agency and Redes Energéticas Nacionais.

Peterson, G.D., G.S. Cumming, and S.R. Carpenter. 2003. Scenario planning: A tool for conservation in an uncertain world. *Conservation Biology* 17: 358–366.

Pierce, J.T. 1992. Progress and the biosphere: The dialectics of sustainable development. *The Canadian Geographer* 36: 306–320.

Pohl, C. 2008. From science to policy through transdisciplinary research. *Environmental Science and Policy* 11: 46–53.

Polèse, M. 1999. From regional development to local development: On the life, death, and rebirth(?) of regional science as a policy relevant science. *Canadian Journal of Regional Science* XXII: 299–314.

Porter, M. 2004. *Competitiveness in rural US regions: Learning and research agenda*. Boston: Institute for Strategy and Competitiveness, Harvard Business School.

Rapport, D., R. Costanza, and A.J. McMichael. 1998. Assessing ecosystem health: Challenges at the interface of social, natural and health sciences. *Trends in Ecology and Evolution* 13: 397–402.

Riesch, H., and C. Potter. 2014. Citizen science as seen by scientists: Methodological, epistemological, and ethical dimensions. *Public Understanding of Science* 23: 107–120.

Ross, W.A. 1998. Cumulative effects assessment: Learning from Canadian case studies. *Impact Assessment and Project Appraisal* 16: 267–276.

Roux, D.J., K.H. Rogers, H.C. Biggs, P.J. Ashton, and A. Sergeant. 2006. Bridging the science–management divide: Moving from unidirectional knowledge transfer to knowledge interfacing and sharing. *Ecology and Society* 11(1): 4.

Salmo Consulting. 2006. *Developing and implementing thresholds in the northwest territories: A discussion paper*. Yellowknife: Environment Canada.

Santomauro, D., C.J. Johnson, and M.P. Fondahl. 2012. Historical-ecological evaluation of the long-term distribution of woodland caribou and moose in central British Columbia. *Ecosphere* 3: 37.

Savoie, D. 1997. *Rethinking Canada's regional development policy: An Atlantic perspective*. Ottawa: Canadian Institute for Research on Regional Development.

Schneider, R.R., J.B. Stelfox, S. Boutin, and S. Wasel. 2003. Managing the cumulative impacts of land uses in the Western Canadian Sedimentary Basin: A modelling approach. *Conservation Ecology* 7(1): 8.

Simberloff, D. 1998. Flagships, umbrellas, and keystones: Is single-species management passé in the landscape era? *Biological Conservation* 83: 247–257.

Steenberg, J.W.N., P.N. Duinker, L.V. Damme, and K. Zielke. 2013. Criteria and indicators of sustainable forest management in a changing climate: An evaluation of Canada's National Framework. *Journal of Sustainable Development* 6: 32–64.

Timoney, K., and P. Lee. 2001. Environmental management in resource-rich Alberta, Canada: First world jurisdiction, third world analogue. *Journal of Environmental Management* 63: 387–405.

Tipa, G., and R.V. Welch. 2006. Co-management of natural resources: Issues of definition from an Indigenous community perspective. *Journal of Applied Behavioural Science* 42: 373–391.

Walters, C. 1986. *Adaptive management of renewable resources*. New York: Macmillan Publishing Company.

Waltner-Toews, D. 2004. *Ecosystem sustainability and health: A practical approach*. Cambridge: Cambridge University Press.

Waltner-Toews, D., J.J. Kay, and N.-M.E. Lister. 2008. *The ecosystem approach: Complexity, uncertainty, and managing for sustainability*. New York: Columbia University Press.

Weber, M., N. Krogman, and T. Antoniuk. 2012. Cumulative effects assessment: Linking social, ecological, and governance dimensions. *Ecology and Society* 17(2): 22.

Wenzel, G.W. 1999. Traditional ecological knowledge and Inuit: Reflections on TEK research and ethics. *Arctic* 52: 113–124.

Westgate, M.J., G.E. Likens, and D.B. Lindemayer. 2013. Adaptive management of biological systems: A review. *Biological Conservation* 158: 128–139.
White, L., and B.F. Noble. 2013. Strategic environmental assessment for sustainability: A review of a decade of academic research. *Environmental Impact Assessment Review* 42: 60–66.
Whitelaw, G., H. Vaughan, B. Craig, and D. Atkinson. 2003. Establishing the Canadian community monitoring network. *Environmental Monitoring and Assessment* 88: 409–418.
Wilcox, B.A., A.A. Aguirre, and P. Horwitz. 2012. Ecohealth: Connecting ecology, health, and sustainability. In *New directions in conservation medicine: Applied cases of ecological health*, ed. A.A. Aguirre, R.S. Ostfeld, and P. Daszak, 17–32. New York: Oxford University Press.

# Author Biographies

**Annie L. Booth**: Annie Booth is a Professor in Environmental Studies in the Ecosystem Science and Management Program at the University of Northern British Columbia. She researches in the areas of Aboriginal perspectives of resources development, community-based resources management, and ethics among others.

**Philip J. Burton**: Phil Burton is a Professor in Ecosystem Science and Management at the University of Northern British Columbia. He currently serves as the Northwest Regional Chair at the UNBC campus in Terrace, BC. With a background in the public service and as a consultant, Phil has undertaken research in various aspects of vegetation dynamics and forest regeneration after disturbance, with a particular interest in ecological resilience and restoration.

**Alana J. Clason**: Alana Clason is a PhD candidate in the Natural Resources and Environmental Studies program at the University of Northern British Columbia. Her research examines northern range dynamics of whitebark pine and the resilience of these populations to changing climate. She is an active board member of the Whitebark Pine Ecosystem Foundation-Canada and an associate researcher with the Bulkley Valley Research Centre.

**David J. Connell**: David Connell is an Associate Professor in Ecosystem Science and Management at the University of Northern British Columbia. He has undergraduate degrees in economics and commerce, a master's degree in business administration, and a doctoral degree that examined the meaning and pursuit of community in modern society. David's current research areas include agricultural land-use planning, farmers markets, and socioeconomic benefits assessments.

**Allan B. Costello**: Allan Costello is an Assistant Professor in the Ecosystem Science and Management Program at the University of Northern British Columbia. His interests lie in the application of science-based research to the management and conservation of native fishes in Canada.



M.P. Gillingham et al. (eds.), *The Integration Imperative*,
DOI 10.1007/978-3-319-22123-6

**Stephen J. Déry**: Stephen Déry has appointments as Professor in the Environmental Science and Engineering undergraduate program and the Natural Resources and Environmental Studies graduate program at the University of Northern British Columbia. His background is in atmospheric science and he has degrees from York and McGill Universities. Stephen also did postdoctoral positions at the Lamont-Doherty Earth Observatory of Columbia University, New York, and held a Visiting Research Scientist position at Princeton University in New Jersey. Stephen is investigating the consequences of climate change on the water cycle of northern and alpine regions, including on snow and ice.

**Michael P. Gillingham**: Mike Gillingham is a Professor in the Ecosystem Science and Management Program and past Director of the Natural Resources and Environmental Studies Institute at the University of Northern British Columbia. His research interests are in population and wildlife ecology, behavioral ecology, and how human use and alteration of landscapes affects wildlife populations. Much of his work focuses on the application of quantitative analyses and modeling to aspects of behavioral ecology and wildlife management.

**Maya K. Gislason**: Maya Gislason is an Assistant Professor in the Faculty of Health Sciences at Simon Fraser University where she co-leads the Social Inequalities in Health stream. Prior to her current position, Maya was a Banting Postdoctoral Fellow at the University of Northern British Columbia. Her research is focused on the intersections of the social and ecological determinants of health, particularly as they play out in relation to social–ecological *game changers* such as climate change and intensive natural resource extraction projects.

**Sybille Haeussler**: Sybille Haeussler is a community-based research scientist and professional forester who lives in Smithers, BC. She is a founding director of the not-for-profit Bulkley Valley Research Centre and an Adjunct Professor in Ecosystem Science and Management at the University of Northern British Columbia. Sybille's work addresses the dynamics, diversity, resilience, and restoration of terrestrial ecosystems and plant communities of northern British Columbia.

**Greg R. Halseth**: Greg Halseth is a Professor in the Geography Program at the University of Northern British Columbia, where he is also the Canada Research Chair in Rural and Small Town Studies and Co-Director of UNBC's Community Development Institute. His research examines rural and small town community development, and local and regional strategies for coping with social and economic change.

**Henry G. Harder**: Henry Harder is an Indigenous Scholar and is Professor and past Chair of the School of Health Sciences at the University of Northern British Columbia. He currently holds the Dr. Donald B. Rix BC Leadership Chair in Aboriginal Environmental Health. His research interests are in Aboriginal health, disability issues, workplace mental health, and suicide prevention.

**Dawn Hemingway**: Dawn Hemingway is an Associate Professor and Chair in the School of Social Work at the University of Northern British Columbia, Adjunct Professor in Gender/Women's Studies and Health Sciences, Co-Director of Women North Network/Northern Fire, and founding member of Stand Up for the North. Her research interests include northern quality of life, aging, and women's health—all informed by empowerment processes such as community-driven research, policy, and planning.

**Derek O. Ingram**: Derek Ingram is a Project Manager and Research Coordinator in Dancing Raven—Communities, Lands, and People, a consulting company in Prince George, BC. He received his MA from the University of Northern British Columbia in First Nations Studies in 2012. His research and community work focuses on First Nations resource management, cultural resource management, and sustainable economic development.

**Peter L. Jackson**: Peter Jackson is a Professor in the Environmental Science and Engineering Programs and the Natural Resources and Environmental Studies Institute at the University of Northern British Columbia. He is an atmospheric scientist whose basic research focuses on the interaction between the atmosphere and Earth's surface over regions of complex terrain. He applies this research to environmental issues such as air pollution, climate change, and avian biology.

**Chris J. Johnson**: Chris Johnson is a Professor in the Ecosystem Science and Management Program and a member of the Natural Resources and Environmental Studies Institute at the University of Northern British Columbia. His research is focused on how human uses of landscapes influence the distribution and population dynamics of terrestrial wildlife. In addition to empirical studies, Chris' research considers the limitations and areas of improvement of policy, legislation, and practice designed to assess and limit cumulative impacts.

**Kathy J. Lewis**: Kathy Lewis is Professor and Chair of the Ecosystem Science and Management Program at the University of Northern British Columbia. She teaches forestry and natural resources management courses and researches the ecological roles of biotic disturbance agents in forests and the influence of management practices and climate change on forest health.

**Nicole M. Lindsay**: Nicole Lindsay is a doctoral candidate in Communication at Simon Fraser University and past research associate at the National Collaborating Centre for Aboriginal Health. Her ongoing research is focused on discourses of sustainability and responsibility in extractive industries and extractives-impacted communities in Canada and Latin America.

**Martha L. P. MacLeod**: Martha MacLeod is Professor and Chair of the School of Nursing and Professor in the School of Health Sciences. She co-leads the University of Northern British Columbia's Health Research Institute. Her research takes a qualitative approach to examining health services and health human resources, particularly in rural and northern settings. She is active in developing national and regional multidisciplinary research and knowledge translation.

**Pouyan Mahboubi**: Pouyan Mahboubi is an instructor of university bio-geo sciences and Dean of Instruction at Northwest Community College. His current research is focussed on approaches to locating important social–ecological spaces in marine ecosystems and examining their implications within the context of the Canadian Environmental Assessment process.

**Indrani Margolin**: Indrani Margolin is an Associate Professor in the School of Social Work, at the University of Northern British Columbia, where she is also a lead researcher for the Centre for Women's Health Research at UNBC. Her research interests include arts-based methodologies, holistic and creative social work practice, and girls' and women's mental health.

**Kendra Mitchell-Foster**: Kendra Mitchell-Foster is a Postdoctoral Research Fellow with the Northern Medical Program and previously with the School of Health Sciences at the University of Northern British Columbia. Her research and work focuses on collaborative strategies to establish and reinforce equity considerations in intersectoral spaces where community-driven evidence and experiential knowledge might influence social and political determinants of health.

**Donald G. Morgan**: Donald Morgan is a Systems Ecologist in the Conservation Science Section of British Columbia's Ministry of Environment. His research develops methods for analyzing and managing socioecological systems with an emphasis on wildlife and salmon. He also leads a collaborative aquatic monitoring program, with the Bulkley Valley Research Centre, where he serves as the Director of Operations.

**Philip M. Mullins**: Phil Mullins is an Associate Professor in Ecosystem Science and Management and Outdoor Recreation and Tourism Management at the University of Northern British Columbia. Phil is an Editor for the *Journal of Experiential Education* and member of the Natural Resource and Environmental Studies Institute. His research examines human–environment relations through participation in outdoor recreation and education activities, and as related to social and environmental sustainability.

**Bram F. Noble**: Bram Noble is a Professor in the Department of Geography and Planning, and School of Environment and Sustainability, at the University of Saskatchewan. His research focuses on environmental assessment, with a particular emphasis on regional cumulative effects assessment and management.

**Katherine L. Parker**: Katherine Parker is a Professor in the Ecosystem Science and Management Program at the University of Northern British Columbia, where she teaches courses in wildlife ecology and animal physiology. She and her students do research that contributes to conservation and resource management, particularly relative to ungulates and their predators on northern landscapes.

**Margot W. Parkes**: Margot Parkes is an Associate Professor in the School of Health Sciences at the University of Northern British Columbia, where she is also a Canada Research Chair in Health, Ecosystems and Society, and is cross appointed with the Northern Medical Program. Margot's research integrates social and ecological determinants of health, focusing especially on watersheds as settings for health, and on education and governance options that foreground the relationships among health, environment, and community.

**Ian M. Picketts**: Ian Picketts teaches at Quest University Canada and is an adjunct faculty member at the University of Northern British Columbia. His primary interests relate to interdisciplinary, action-oriented research focusing on climate change adaptation and sustainable resource management.

**Justina C. Ray**: Justina Ray is President and Senior Scientist of Wildlife Conservation Society Canada. She is involved in research and policy activities associated with conservation planning in northern landscapes, with a particular focus on wolverine and caribou. She also serves as Adjunct Professor at the University of Toronto (Faculty of Forestry) and Trent University (Biology Department).

**Glen Schmidt**: Glen Schmidt is a Professor in the School of Social Work at the University of Northern British Columbia. He practised social work in northern and remote resource-dependent communities and is interested in social service delivery in this context.

**Dale Seip**: Dale Seip is a wildlife ecologist with the BC Ministry of Environment in Prince George, BC. He has been involved in caribou research in BC for over 30 years. He has also been involved in recovery and management planning for caribou at the national, provincial, and local level.

**Karyn Sharp**: Karyn Sharp is Denésuliné from northern Saskatchewan and is a partner and Project Coordinator in Dancing Raven—Communities, Lands, and People, a consulting company based in Prince George. She is also an Adjunct Professor in Anthropology at the University of Northern British Columbia. Her research and community-based work focuses on First Nations governance, traditional food systems, and cultural resource and land management.

**Nobuya (Nobi) Suzuki**: Nobuya (Nobi) Suzuki is a Research Associate in the Natural Resources and Environmental Studies Institute at the University of Northern British Columbia. His recent research examines conservation of wildlife and biodiversity in relation to potential natural resource management activities in northeast British Columbia.

**Cathy Ulrich**: Cathy Ulrich is President and CEO of Northern Health (NH). From 2002–2007, she was NH's Vice-President of clinical services and chief nursing officer. Before the formation of NH, she worked in a variety of nursing and management positions in northern British Columbia, Manitoba, and Alberta. She spent the majority of her career in rural and northern locations, where she gained a solid understanding of the nature of local communities, their health needs and concerns, and the unique approaches required to meet these needs. Ulrich is an Adjunct Professor at the University of Northern British Columbia's School of Nursing and School of Health Sciences.

**Pamela A. Wright**: Pam Wright is Associate Professor in the Outdoor Recreation and Tourism Management Program/Ecosystem Science and Management at the University of Northern British Columbia. Her research focuses on the ecological management of parks and the connections between health, nature, and the outdoors.

# Index




M.P. Gillingham et al. (eds.), *The Integration Imperative*,
DOI 10.1007/978-3-319-22123-6

CPSIA information can be obtained
at www.ICGtesting.com
Printed in the USA
LVOW02*1331090416
482895LV00003B/21/P